Rajesh Chandra

Industrialization and Development in the Third World

London and New York

First published 1992
by Routledge
11 New Fetter Lane, London EC4P 4EE

Simultaneously published in the USA and Canada
by Routledge
a division of Routledge, Chapman and Hall, Inc.
29 West 35th Street, New York, NY 10001

© 1992 Rajesh Chandra
Typeset by J&L Composition Ltd, Filey, North Yorkshire
Printed and bound in Great Britain by
Biddles Ltd, Guildford and King's Lynn

British Library Cataloguing in Publication Data
Chandra, Rajesh
  Industrialization and development in the Third World.
  I. Title
  338.091724

ISBN 0–415–01380–1

Library of Congress Cataloging in Publication Data
Chandra, Rajesh.
    Industrialization and development in the Third World / Rajesh
  Chandra.
        p.   cm.—(Routledge introductions to development)
    Includes bibliographical references and index.
    ISBN 0–415–01380–1
    1. Developing countries—Industries.   2. Developing countries—
  Dependency on foreign countries.   3. Industry and state—Developing
  countries.   I. Title.   II. Series.
  HC59.7.C339   1992   91–30175
  338.09172′4—dc20   CIP

Contents

Plates

Figures

x Figures

Tables

Acknowledgements

I am most grateful to the Series Editors, John Bale and David Drakakis-Smith, for asking me to prepare this book, and for their editorial assistance and patience during its preparation.

Portions of this book were completed while I was a Visiting Professor of Geography at McGill University, Montreal, Canada, and Visiting Research Fellow, Macmillan Brown Centre for Pacific Studies, University of Canterbury, Christchurch, New Zealand. I would like to thank Professor Sherry Olson, Chairman of the Department of Geography at McGill University, and Dr Malama Meleisea, the Director of the Macmillan Brown Centre for Pacific Studies, for their assistance. Most of the cartographic work for the book was undertaken by Mr Tony Shatford of the Geography Department of the University of Canterbury, to whom I am very grateful. Mr Ariel Prakash of the Editorial Office of the School of Social and Economic Development, University of the South Pacific, and Mr Andrew Lawrence of the Department of Geography at the University of Keele drew the remaining diagrams. I would also like to thank the British Council whose visiting programme has assisted in the preparation of the book. Finally, my wife has been of invaluable help, as always, during this project, and I am most grateful for her support.

Preface

Most Third World countries (the term is used in this book synonymously with the term 'developing countries') are committed to transforming or changing their rural-based agricultural economies to urban-based industrial ones. There may be differences in the level of industrialization they wish to achieve, the speed at which they wish to industrialize, or in their industrialization strategies, but nearly all of them are strongly committed to their goal of industrialization.

This commitment to industrialization is not confined to large developing countries. Some small countries, such as Hong Kong, Singapore and Taiwan, have achieved a significant degree of industrialization. Even small countries in the South Pacific, such as Fiji, are committed to the same route.

Developing countries desire industrialization because they realize that it is inextricably linked to development and that, historically, industrialization has been the only path to development. Countries that are classed today as developed have all gone through an industrial revolution. Indeed, there is almost no country that one would be prepared to call 'developed' that has not gone through an industrial transformation.

Principal themes

Industrialization and development in developing countries can be studied in many ways by using different perspectives and by studying varied topics, or aspects. We can best conceptualize Third World industrialization as depicted in the figure on p. xvi, which identifies the key

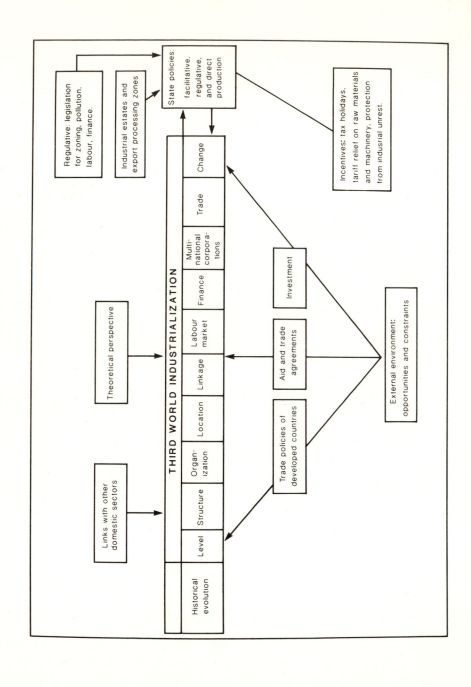

areas usually studied and the main influences on and issues in the manufacturing sector.

Given the introductory nature of this book, we will concentrate only on a few main aspects of industrialization and development in the Third World. Chapter 1 explores the relationship between industrialization and development and introduces the main concepts used in the book. Since conflicting interpretations of Third World industrialization arise from the use of different frameworks or perspectives, these are introduced as well. Chapter 1 also emphasizes the critical role of the external environment in Third World industrialization. Chapter 2 then examines the historical dimensions of this industrialization, arguing that the way Third World countries were absorbed into the international economy has been a major influence on their current level and structure of industrialization.

Chapter 3 describes the level and structure of industrialization in developing countries, in particular its unevenness, and the way it has concentrated in the past mainly on food and light labour-intensive industries. However, in recent years, developing countries have been creating heavy and sophisticated industries, with considerable success. But much of the success of the developing world in manufacturing has been confined to what are called newly industrializing countries (NICs), such as South Korea or Taiwan.

Industrial organization in the Third World is the subject of Chapter 4, which examines issues of ownership (foreign and local; private and public) and forms of commercial organization, focusing on small-scale manufacturing. The location of manufacturing production is discussed in Chapter 5. Since there has been a great deal of interest internationally in the movement of manufacturing investment offshore from developed to developing countries, this chapter explores both inter-country location and intra-country locational patterns.

The state has become more and more closely involved in the economies of the Third World, and its influence on industrialization is especially marked. Chapter 6 examines this issue, as well as reporting on the main industrialization strategies followed by the various governments. Finally, Chapter 7 synthesizes the main discussions of the book and assesses the future prospects of Third World industrialization.

1
Industrialization and development in the Third World

Introduction

The relationship between industrialization and development is surprisingly varied and many reasons have been put forward to explain why Third World countries are so committed to industrialization. Some of the principal arguments are as follows:

1 Industrialization is seen in the Third World as necessary because of its historical association with development. Because of the absence of any other demonstrable model of development, historically, it is taken for granted that development entails industrialization.
2 Industrialization is also favoured by developing countries because they have exhausted the possibilities of agricultural development and because prices of agricultural products have fluctuated wildly in the past. These prices have also not kept pace with the prices of manufactured goods. In other words, the terms of trade for agricultural commodities have deteriorated.
3 In addition, as incomes increase, there is no proportionate increase in the consumption of agricultural products; that is, the income demand elasticity of agricultural products is low, reducing its long-term developmental potential. On the other hand, manufactured goods have a higher income demand elasticity. Moreover, many agricultural products are facing major problems of reduced consumption due to changes in lifestyle such as reduced consumption of sugar and related products, or from the rise of synthetic products.

4 Even when manufacturing is not seen as an alternative to agricultural development, it is encouraged because it complements the agricultural sector. Most developing countries are agricultural societies. The development of manufacturing can help the agricultural sector in many ways. The processing of agricultural commodities, which is part of manufacturing, increases the income of a country because the more processed a commodity is, the higher is its value. The United Nations Conference on Trade and Development has estimated that further processing of agricultural commodities exported by developing countries could increase their income by at least 50 per cent.

Manufacturing also encourages efficient forms of production and marketing in the agricultural sector, provides agricultural inputs such as machinery and fertilizer, and improves the availability of food items by making them available as processed foods. Food processing can also eliminate the problem of market glut by providing an outlet for excess production. Furthermore, increased industrialization can improve the bargaining position of regional states and national governments because processing makes commodities less perishable. Manufacturing can also help the agricultural sector by absorbing labour from the rural sector, thus enabling the mechanization and rationalization of agriculture. A degree of mechanization is essential for increased productivity in the agricultural sector.

5 The populations of most developing countries are increasing rapidly (see, for instance, the book by Allan and Anne Findlay in this series). Employment generation has not kept pace with population growth, and unemployment and underemployment are high and increasing. Manufacturing has been seen as a major source of additional employment. This is especially so as the traditional sources of employment, such as agriculture, mining, services and construction, have become employment saturated. Manufacturing does provide for a reasonably high proportion of the employed labour force in many developing countries, but many critics of Third World industrialization policies have argued successfully that the highly capital-intensive nature of industrialization has diminished its contribution to the reduction of unemployment and underemployment.

6 Manufacturing is also favoured as a development strategy because of its efficient use of land resources. Agriculture is an extensive user of land, which is a finite quantity. Indeed, the amount of land available to a society can and does decrease with time as more and more of it is lost to deserts or becomes only marginally productive.

Manufacturing becomes attractive because of its more efficient use of land. Thus, for small countries, such as Hong Kong and Singapore, there was no alternative but to industrialize.

7 One of the important development goals of developing countries is to evolve into integrated societies both economically and spatially. A society with a sense of shared identity, and one closely knit together, is more likely to succeed in development than one without these attributes. Industrialization promotes national integration. Manufacturing involves a large number of transactions (farmers selling raw materials to wholesalers or manufacturers, who sell to wholesalers after processing; manufacturers purchasing electricity, legal services, communications and so on) both within the country and outside it, which help to develop stronger and greater links. The greater the degree of linkage, the greater is the interdependence and the possibility of building a spatially integrated society.

8 The initial justification for industrialization in many instances was that it would save foreign exchange by producing what was previously imported. In fact most developing countries began their industrialization through this import-substitution strategy. However, in the final analysis, the savings in foreign exchange were limited or absent, since foreign exchange was spent on machinery imports, licence fees and the import of raw materials instead of on the import of finished manufactured goods, as happened previously.

Industrialization was also seen as having the potential to earn foreign exchange with exports after entrepreneurs had acquired the necessary expertise and met domestic demand. In some cases this did indeed happen; in others it did not, because most manufacturers were satisfied with the returns they made on the domestic market and because their goods, in any case, were not competitive in regional and international markets in terms of price or quality.

9 Many Third World governments pursue industrialization because they wish to reduce their technological dependence on the developed countries. Technology is the chief basis of economic production and, in particular, of increasing productivity. To some extent, some Third World countries have indeed developed considerable technological capacity; in many other cases, however, industrialization has made these countries more dependent on developed countries by locking them into technology licensing agreements and debt relationships.

10 The most powerful countries of the world are also the most

industrialized. This is no coincidence – the two are closely related. Many Third World countries, such as Brazil, India, China and Israel, are all conscious of the military role of industrialization. This is especially borne out by their commitment to large-scale heavy industry and now, increasingly, to the development of advanced electronics.

It is for these varying reasons that industrialization has such appeal and elicits such strong commitment from Third World planners and politicians.

Definitions

Since this book revolves around a few main concepts, it will be useful to explain at the outset what they mean. Some broader definitions are discussed in Case study A, pp. 7–9.

Industrialization refers to an increase in the share of the gross domestic product (GDP) contributed by the manufacturing sector. It is a process that involves a change in the structure, or make-up, of the economy. Industrial growth in itself is not sufficient for industrialization, because other sectors of the economy may increase their output at the same rate. It is necessary for the manufacturing sector to increase its relative importance in the economy more rapidly than other sectors. It is important to keep this point in mind because of the frequent confusion between industrial growth and industrialization.

The composition of an economy is measured by broad groups of economic activity using the International Standard Industrial Classification of All Economic Activities (ISIC). The main ISIC groups are: 1 Agriculture, fisheries, forestry; 2 Mining and quarrying, 3 Manufacturing; 4 Utilities; 5 Construction; 6 Wholesale, retail, restaurants and hotels; 7 Transport and communications; 8 Finance, insurance and real estate; 9 Community and personal services; 10 Activities not elsewhere classified.

The above definition of industrialization should not give the impression that it is a purely economic process; it is much more than that. The process of industrialization is, in its broadest sense, a process of societal transformation, involving economic, political, social and cultural changes. Industrialization implies greater economic specialization in production geared to national and international markets and a significant increase in the share of manufacturing in the total output of a country and in the absorption of resources. It also implies the use

Figure 1.1 Implications of industrialization

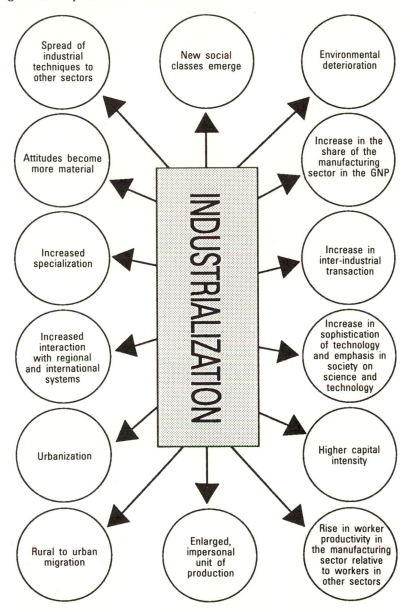

of science and technology in production, leading to the production of goods based on complex technology and capital-intensive techniques; changes in work organization leading to scientific management and increased productivity; the spread of industrial techniques to the rest of the economy; a shift in attitudes and relationships to material phenomena; and a shift to urban settlement. Figure 1.1 provides more details of the various attributes of industrialization, underlining the very broad nature of the process of industrialization.

Another term that will be used extensively in this book is *industrial*. This is usually taken to mean economic activities covered by ISIC groups 2 (mining and quarrying), 3 (manufacturing) and 4 (utilities). The censuses of industries internationally include all these groups, although they report on them separately.

Manufacturing is a subgroup within the industrial category. It refers to activities that transform or combine materials into new products to make them more valuable (in terms of money earned from them) or more useful. Manufacturing is listed under ISIC group 3 and includes both factory- and non-factory-based activities. As we will see later in the book, a large proportion of manufacturing output in developing countries is produced in the informal sector, or cottage industry. Even in developed countries, the importance of home-based manufacturing output is increasing, especially with improved communications technology.

Finally, the term *value added* will be used frequently throughout the book. In order to avoid duplicate counting of the contribution of various kinds of economic activities, only the additional value created by an activity is counted. This means that even when one activity provides materials or services for another activity, such as a steel mill providing raw materials for shipbuilding, no duplication in counting its contribution occurs; only the additional value is counted. Value added refers to the difference between the gross output of an activity (roughly equivalent to gross sales) and the cost of producing the item (such as raw materials, electricity, finance charges, advertising and other miscellaneous expenses, but excluding labour costs). Labour costs are therefore included in value added which is the equivalent of gross profits and labour charges.

Case study A

Third World dependence

The Third World faces a major obstacle in its attempts to industrialize because of its acute dependence on developed countries in most major areas. This dependence can be seen in technology, trade, foreign investment, human resources, military hardware, aid, and information flows and technology.

Technology

Ever since the industrial revolution in Europe in the nineteenth century, the Third World has been dependent on the developed world for most of its technology. Even today, although some developing countries possess an ability to supply technology to other developing countries, the Third World still relies on developed countries for most of its technology, especially for leading-edge technology. Ironically, since most of the machinery has been developed in developed countries it does not always suit developing countries. Machines are capital intensive, reducing employment, when in developing countries there is a need to use more labour, not less.

Markets

Most developing countries export the bulk of their manufactured goods to developed countries, particularly to the United States of America and the European Community. However, developing country exports can easily be affected as developed countries slowly close their markets to imports from developing countries because of domestic unemployment and pressure from domestic manufacturers and politicians.

Foreign investment

Although the Third World relies mainly on its own resources for development, it needs foreign investment to speed up its development and access to new technology, and to gain new markets often represented by multinational corporations. This dependence is not

Case study A *(continued)*

without problems. Multinational corporations frequently displace national corporations and exert extensive economic and political influence. Many small developing countries are not able to deal adequately with giant multinational corporations.

Human resources

Many Third World countries still depend on developed countries for the training of their nationals and for experts. Third World education is greatly affected by developed countries because of the frequent use of theories and models developed in these countries which have vastly different socio-economic, political and cultural environments. Furthermore, many Third World nationals are trained in developed countries, and many developed country experts are used as teachers and policy advisors in developing countries.

This educational dependence harms Third World countries by making it more difficult for them to develop appropriate theories and models, and to formulate government policies that are both appropriate and in their interest.

Military dependence

It goes without saying that military weakness is implied in the Third World status of countries. The Third World, on the whole, depends on developed countries, both capitalist and socialist, for its security arrangements and for military hardware. While some developing countries, such as China, India, Israel and Brazil, have become significant arms exporters, the Third World still relies heavily on developed countries for most of its arms, especially state-of-the-art military hardware. This military dependence means, among other things, that developing countries must align themselves with their security partners on regional and global security and general affairs, even if these are not in their interest.

Case study A *(continued)*

Aid

While Third World development is financed mostly from domestic resources, foreign aid has become crucial to many countries as an added source of foreign exchange, human resources and technical assistance. In some cases, aid is needed to deal with natural and man-made disasters, such as hurricanes, droughts and civil war. However, aid is often tied, meaning that goods and services from the donor countries must be used to a large extent. This makes a recipient country dependent on donor countries for future services and technology, often at a higher cost than that available on the open market. In addition, aid is often used to induce changes in the domestic and foreign policies of countries.

Information technology and flows

The dependence of the Third World on developed countries is felt sharply in the area of information technology. This dependence is extremely important because of the very strong and direct link information technology has with all facets of a country's performance.

Developing countries depend on developed countries for the latest technology in satellite communications, computing hardware and software, and for the actual gathering, storage and dissemination of news. This developed country control of news, together with control of the leading communications technology, means that developing countries are at a major disadvantage in nearly every facet of their activities.

The international context of Third World industrialization

It is now increasingly realized that the international economic system strongly influences the industrial progress of developing countries. Although this view has been expounded largely by the contentious school of thought known as the dependency school, nearly all development writers now consider the international environment an important factor in the industrial progress of developing countries.

The international environment affects the industrialization of developing countries both directly and indirectly (Figure 1.2). Direct influences are felt through the trade practices of developed countries, particularly their tendency to protect their own domestic manufacturers through the imposition of tariffs. Since most developing countries rely on markets in developed countries for their success, the increasing

Figure 1.2 The international economy and the Third World

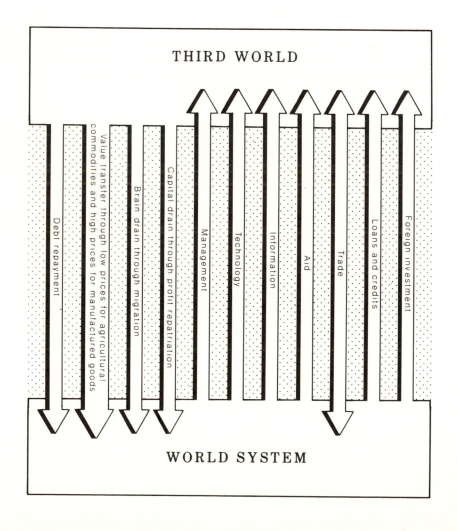

protectionism observed in many developed countries poses a real threat to them.

Protection of domestic manufacturing production takes many forms, some direct and easily measurable, some subtle and more difficult to measure. The main forms of protection are through tariffs, which artificially increase the price of imported goods, and quotas and licences, which aim to regulate the flow and volume of imported goods. Some countries also take more subtle protectionist measures, such as threats to impose restrictions unless other countries agree to voluntary restraint. The United States of America has used this technique effectively against imports from Japan. Countries sometimes also use bureaucratic red tape to protect domestic industries, such as the imposition of excessively rigorous quality standards and checks. Many western countries accuse Japan of using this technique to discourage imports.

Direct influences of the international environment on Third World manufacturing are also felt by the presence of multinational corporations (MNCs) in developing countries. MNCs are extremely large global organizations and their operations in developing countries have major consequences for them, especially given their low level of development. For instance, many MNCs have been known to affect national policies relating to industrialization through political lobbying and/or through pressures applied by their parent countries.

Aid and trade agreements also affect Third World industrialization, both directly and indirectly. Many developing countries, especially small countries, receive aid specifically for industrial projects. The European Community, for instance, has been assisting the small island countries of the South Pacific with personnel experienced in the management of industrial operations, arranged through the Centre for Development of Industry based in Brussels. Australia and New Zealand have also helped through aid projects, and through a trade and aid agreement called the South Pacific Regional Trade and Economic Cooperation Agreement (SPARTECA).

Indirect influences can be seen in the case of prices paid for Third World commodities. When commodity prices are high, the internal economy of the Third World becomes buoyant, and internal markets for manufactured goods expand. Government revenue also increases, making it possible for it to provide more infrastructural support to the manufacturing sector. When commodity prices are low, the whole economy sags, and there is no expansion in the domestic market;

government revenues decline and are frequently prioritized for the agricultural and rural sectors.

Foreign debt too influences industrialization because of its crippling effects on Third World economies. Indebtedness has increased dramatically in the last decade to historically unprecedented levels. Many Third World countries are facing extreme difficulty in meeting their debt charges. While the more prosperous developing countries can probably pay their debts in the future, the poorer countries, especially in Africa, cannot.

Given such extreme pressure to pay debts, there is a tendency to concentrate on agricultural development, where exports are expected to help raise necessary revenues. Import-substituting industrialization, which serves the local market, is discouraged and, in some cases, disbanded. Emphasis is placed on those types of manufacturing that will earn foreign exchange to satisfy foreign creditors.

So, for these reasons, the external environment of developing countries represents both constraints on and opportunities for the expansion of their industrial production.

Plate 1.1 The Guangzhou Peugeot Automobile Company, a Sino-French joint venture in China. China has long used the joint-venture form of foreign investment, although since the 1970s it has adopted a more open-door policy, allowing majority foreign-owned firms to be set up.

Viewing Third World industrialization: different perspectives

The study of industrialization in developing countries can be undertaken from different perspectives. These different approaches greatly affect the conclusions drawn as well as the nature of the studies themselves. The use of different perspectives has largely been responsible for the conflicting interpretations of industrialization in the Third World, seen most dramatically in the works of André Gunder Frank and Bill Warren.

Four principal perspectives can be identified: studies in industrial geography; the neoclassical economic approach; Marxist approaches, including both orthodox and neo-Marxist schools; and, finally, dependency approaches. These are briefly reviewed below.

Industrial geography

Industrial geography was so completely dominated by industrial locational analysis and largely on the thinking of Alfred Weber in the past that the term 'industrial geography' is probably a misnomer as a description of much of the early work of this approach. The preoccupation with locational analysis with all its shortcomings – to the exclusion of other equally important, if not more important, topics – has severely limited the contribution of industrial geography to an understanding of industrialization in both developed and developing countries.

Industrial geographers have been mainly interested in identifying the location of manufacturing establishments and in developing models to explain these locations as well as to indicate idealized locations to minimize operational costs and maximize revenue. In many instances, however, people working on locational theory were interested in models of what location should be, rather than what the location was. Many of the models depicting idealized locations were based on very unrealistic assumptions about human behaviour, so much so that these models bore little resemblance to reality.

Furthermore, location is only one aspect of industrialization. Many writers have argued that for the vast majority of small manufacturers decisions regarding location are subordinate to other business decisions. In some cases, manufacturers take it for granted that they will locate where they happen to be living.

As we have already seen in the figure in the Preface, industrialization is a broad process and studying only part of it (its location) gave a limited understanding of this crucial process. Industrial geographers did realize these shortcomings and, at least from the late 1970s, have considerably broadened their scope to include the study of the whole industrial system. They have also begun to pay more attention to the study of processes that bring about patterns than to the patterns themselves.

Neoclassical economic approach

In the studies of industrialization in developing countries neoclassical economists are dominant. Studies such as those by the OECD already referred to, and more recent studies, such as the book edited by Cody, Hughes and Wall, *Industrial Policies for Developing Countries*, illustrate well the main thrust of the neoclassical approach to Third World industrialization. A central feature of this work is that it comprises full-scale studies of the industrial systems of countries (in sharp contrast with work by industrial geographers). Topics most frequently covered by neoclassical economists include industrial structure, utilization of factors of production, types and levels of protection, public industrial policies, financing, industrial technology transfers, income distribution, multinational corporations, and trade policies.

However, despite its pre-eminent position, and advanced tools, neoclassical economics has not been able to make a major contribution to our understanding of Third World industrialization because of its underlying assumptions, especially those relating to the effectiveness of the market mechanism, the mobility of factors of production and the role of the state. Furthermore, neoclassical economics has promoted the concept of value-free social science and has elevated essentially social issues to the status of technical problems. Many of these studies also suffered from a complete neglect of the historical dimension of contemporary industrialization.

Marxism

Marxism has increasingly challenged the conventional ways of looking at socio-economic issues in the last two decades. The rise of Marxism has coincided with the failure of conventional theories to deal with rising unemployment, socio-economic inequality, environmental degradation and social alienation. Marxism has a thoroughgoing historical approach to the questions it addresses. This comes, of course, from the emphasis Marx himself placed on historical analysis. Marxism also emphasizes the holistic approach – in seeing problems in their totality, in contrast to neoclassical economics' tendency to carve out specific problems. This emphasis stems from the Marxist belief in the interrelatedness of things.

Finally, Marxism emphasizes the search for the underlying, fundamental explanations.

In addition to these methodological aspects, Marxism overcomes one of the major deficiencies of conventional social science by its emphasis on class and power analysis. Clearly power relations are critical in any social interaction; and classes represent the fundamental groups in society – even if we do not accept bipolar categorization of society into essentially workers versus capitalists. And finally, Marxism has a strong interest in changing the situation rather than studying for studying's sake. This idea of change, and of practice, is firmly ingrained in Marxism.

Dependency

The dependency perspective consists of two main components – a methodology for undertaking studies of the Third World, and a set of ideas, not quite a theory, that guides explanation. It is also important to note that many of the explanations used by this perspective draw liberally from Marxism.

Dependency writers insist, like Marxists, on a thoroughgoing historical analysis of situations, arguing that part of the explanation of the underdevelopment of the Third World is to be found in the nature of the contact between it and the now developed countries. Dependency writers also insist on a holistic approach to Third World development, holding that one cannot analyse individually, and in isolation, the various components of what is a thoroughly linked system. Thus, international trade is frequently considered in terms of international power relationships, and industrialization is considered in the context of the overall development strategies of the Third World. The dependency perspective also rejects the view of modernization theorists, and also of orthodox Marxists, that the causes of Third World development are to be found in the internal structures of these societies. They maintain that the external relationships of countries themselves have to be analysed, not only to examine why they have evolved in the way they have, but to evaluate their future development strategies.

According to the dependency perspective, Third World societies are peripheral and dependent. Peripheral societies are those where the development of capitalism has been blocked, and thus prospects of development reduced. Samir Amin, who has written extensively on the subject, bases his discussion of the peripheral social formation on a distinction between the accumulation process in the central (core) countries and that of the periphery. His basic argument is that in the case of the central (core) economies, development was based on the capital goods sector and the production of goods for mass consumption. In these societies, wages were both a cost and a demand for (further) production. These societies did not depend on external relations to any significant degree. He argues that the sequence of production – the early appearance of mass production, the early transformation of agriculture – was based on demand in urban areas and the late arrival of consumer durables.

In the peripheral economies, however, a 'distorted' picture allegedly prevails. The process of accumulation began with outside impulses. Wages were always seen more as costs than components of demand, and were, therefore, kept low, preventing the acceleration of demand for manufactured goods which had happened in the core countries.

Because of the limitations imposed by adopting any particular perspective on Third World industrialization, this book will use concepts and models from these different perspectives as and when appropriate. The book will be guided mainly, however, by what is actually happening in Third World countries as they industrialize; it will not try to test any theory.

Key ideas

1 Most Third World countries are committed to transforming or changing their rural, agricultural societies to urban, industrialized economies.
2 Industrialization and development are inextricably linked and governments hope that industrialization will earn foreign exchange, provide employment, spatially integrate the economy, and reduce Third World technological dependence on developed countries.
3 Industrialization is a process of change in society in all its aspects; it is not just an economic process. Industrialization leads to the manufacturing sector becoming relatively more important in the economy than other sectors; to the appearance of a working class; to the concentration of settlement in urban centres; and to the rise of the importance of science and technology.
4 The external or world environment presents both significant advantages, such as markets, capital or technology, and imposes constraints, such as increasing protectionism, on Third World development.
5 There are conflicting interpretations of the success of industrialization in developing countries, based largely on the use of different theoretical concepts.

2
The historical dimensions of Third World industrialization

Introduction

The current economic, social and political situation of developing countries cannot be properly understood without an adequate understanding of its historical background. Contemporary successes and failures of the Third World are deeply rooted in the historical evolution of these societies. Furthermore, the forces that have shaped the Third World are still active, and they are a powerful impediment to its fuller development.

A historical appreciation of the Third World is especially important because the great chasm that divides it from the more wealthy, developed and industrialized countries is a relatively recent phenomenon. Many students, not exposed to history, are apt to forget this as they develop stereotypes of the Third World as poor, technologically backward and culturally inferior societies. The truth is that, until the onslaught of European imperialism, these societies were culturally vibrant, economically wealthy and technologically advanced, in many cases more advanced than Europe itself. Specific cases can be singled out: the Aztec and Inca civilizations of the Americas, India, China, Egypt, and the Muslim civilization that stretched from southern Spain to the Indian Ocean.

While Europe was far behind some of these societies, it was Europe that evolved from feudalism to capitalism. This development was nascent in China and India, but these societies were never able fully to evolve to capitalism. The evolution of capitalism in Europe enabled it to develop an industrial, technological and intellectual base that was lacking in other societies. Capitalism also led to the industrial revolution in Britain in the late eighteenth and early nineteenth centuries, and this industrial revolution quickly spread to other countries, such as France, Germany and the United States. The industrial revolution needed overseas outlets for investment, overseas sources for its raw materials and overseas markets for its manufactured goods, and led to the overseas expansion of Europe. At the same time, this overseas expansion contributed to the success of the industrial revolution itself.

Colonialism

It is also important to realize that empires predate European colonialism, for there were Indian, Chinese, Inca and Arabian empires well before those of European countries. European empires differed from these early empires in that they were pervasive in their influence and created a world system with Europe as its centre. It was a world system from which few countries could escape; European empires engulfed most of the world, with the exception of only a few countries such as China or Thailand. All of these empires functioned as part of one world system. Even when formal colonies were not established, societies were opened up for international trade, often against their will and certainly against their interests. China is a case in point, with Britain establishing a trade in opium after the Opium Wars of 1839–42 to finance its trade imbalance with China. The colonies were inducted into the world system in the interest of the colonial powers and at a pace determined by them.

The British geographer Dickenson (1983) has divided European expansion into two phases, the first phase being from 1450 to 1800 and the second phase from 1800 to 1945.

Phase 1: 1450–1800

The first phase of European expansion was largely confined to the Americas, first Central and South America and then North America. It involved conquest, plunder and some settlement by European people. During this phase, Europe was not the most powerful region in the world, for as we have seen there were other empires, such as the Chinese, Indian and Arabian. However, it was Europe that improved its marine technology and was inclined to conquer and settle other lands.

Colonialism in the Americas involved the Spanish, Portuguese,

French, Dutch and the British. Spain was by far the most important of the Latin American colonial powers, followed by Portugal. Initially, the Pope granted exclusive rights to the New World, as the Americas were to be known, to Spain, but the Treaty of Tordesillas in 1494 divided it between Spain and Portugal. This act showed both how important these two powers were then, and the critical role of the Church in colonialism. The other powers came later, and colonized other areas, notably the British in the New World and in the Caribbean, and the French and British in North America.

This first phase of colonialism can be called mercantile capitalism, when the interests of the powers were mainly to plunder, settle and trade, but not to involve these colonies in the extensive production of raw materials, or to use them as markets. Trade interests were predominant, despite the fact that during this period Europe did not have much that it could have sold to these societies in return for the valuable materials it wanted.

During this phase, there was considerable settlement of Europeans in some parts of these regions, particularly those with a similar climate to that of Europe. Spain and Portugal settled South American countries, starting as early as the beginning of the sixteenth century; and the British and French settled in North America. In addition, the slave trade began in this period, slaves being needed to work the mines in Latin America and to produce traded goods, such as sugar and later cotton. These developments were to have an impact on the development of the colonies and on their industrialization, as we shall see later in this chapter.

In the nineteenth century, Britain experienced far-reaching social, political, economic and financial changes that brought about the industrial revolution, which then itself led to further changes. Towards the end of the nineteenth century, despite British efforts to prevent it, the industrial revolution spread to the rest of Europe, particularly to Germany, which had been united since 1871. The industrial revolution needed large quantities of raw materials for mass production and large, preferably protected, markets for mass-produced goods, leading to a quickened pace of colonialism involving a larger number of countries. Economic exploitation rather than settlement was the driving force of the second phase of European expansion.

Phase 2: 1800–1945

In marked contrast with the mercantile period, expansion during this phase was extremely rapid. For instance, whereas in 1714 European

Table 2.1 British colonial expansion, 1814–1906

Date	Country
1814	British Guiana
1816	Gambia, Sikkim
1819	Singapore
1821	The Gold Coast
1826	Assam
1833	Falkland Islands
1839	Aden
1840	New Zealand
1841	Hong Kong
1842	Natal, Sind
1846	North Borneo
1849	Punjab
1852	Burma
1853	Nagpur
1854	Baluchistan
1861	Nigeria
1868	Basutoland
1874	Fiji
1878	Cyprus
1882	Egypt
1884	Somaliland
1887	Zululand
1888	Southern Rhodesia, Sarawak
1890	Kenya, Zanzibar
1891	Northern Rhodesia, Nyasaland
1894	Uganda
1900	Transvaal, Orange Free State
1906	Swaziland

Source: Greene, Felix (1970) *The Enemy*, London: Jonathan Cape.

possessions covered only 10 per cent of the world's land area and less than 2 per cent of the population, by 1914 these had increased to 56 per cent and 34 per cent respectively. Some impression of the rapidity and the scale of colonialism in this period can be obtained by examining the growth of British colonialism from 1814 to 1906, presented in Table 2.1

The second phase involved industrial capital rather than mercantile capital. This means that there was much more interest in the production of raw materials, rather than simply the right to trade in the goods already produced by the Third World, such as tea and spices. Moreover, there was much greater interest in the control of these areas for markets, and to use these markets against the interests of other competing countries. By this time, world leadership had already passed from Spain and Portugal to Britain, which had fought major wars with France to establish its hegemony: thus Britain was the overwhelmingly important colonial power in this phase. At the height of its colonial power, Britain controlled one-fifth of the world's land and a quarter of the world's population. This phase of European expansion is also marked by the participation of other powers as well, which is seen best in the scramble for Africa, when the later emerging industrial powers, such as Germany and Italy, also claimed colonies. However, most of Africa was shared between Britain, which took large territories in the east and the south, and France, which was active in the west and the north of the continent (see Case study B, pp. 24–6).

Decolonization and independence

If the second phase of European colonialism was rapid, the process of decolonization was even more so, especially after the Second World War. This process of decolonization had started early, even as some other parts of the world were being colonized. In 1822, for instance, Brazil had secured its independence. At this time, of course, India was still to be fully colonized; this was achieved only in 1858.

The Latin American countries led the way in decolonization. In many of the countries independence was achieved after a long and hard armed struggle, while in others, especially where the colonial rule was not very harsh, the transition was much more smooth. Most countries, however, achieved independence only after the conclusion of the Second World War. With the major colonial powers reduced to a secondary role after the rise of the United States to world ascendancy, the colonies were given independence quickly. Clearly the granting of independence to India in 1949, after a long and difficult struggle, led to the ultimate dismantling of the British Empire. Many other areas drove out their colonial masters, as happened in Indo-China and parts of Africa.

One of the major contributing factors to independence after the Second World War was the fact that many of the leading colonial powers had been devastated economically and the United States had risen as a world power..It was in the interest of the United States to promote the independence of countries, for this would open these countries to its influence and economic penetration. Of course, the United States was also led by its zeal for independence, something it had fought for and consequently valued highly.

The legacy of colonialism

There is considerable debate among academics about the impact of colonialism. To some extent, an assessment of the colonial impact in the Third World depends on the perspective that one adopts, just as the adoption of different perspectives leads to different perceptions of the nature and level of industrialization in the Third World. To illustrate, many liberal apologists of the colonial era refer to the many benefits that the colonies derived from colonialism, such as the introduction of law and order, educational systems and the development of physical infrastructure (roads, railways, ports). Classical Marxists also saw colonialism positively, because they argued that colonialism led to the destruction of precapitalist modes of production, and to the introduction of capitalism, which they regarded as a dynamic mode of production, one that prepares the ground for eventual socialist revolution. This view has been challenged by neo-Marxists and dependency writers, most of whom argue that colonialism represented an exploitation of the colonies, and that capitalism, even when introduced, was of a different nature and could not lead to the development of these societies. Whatever perspective is adopted, several observations can be made on the impact of colonialism.

1 There is little doubt that colonies were deprived of valuable mineral, agricultural and other resources. To give but three examples: between 1600 and 1810, at least 22,000 tons of silver and 185 tons of gold were transferred to Spain from South America; in only half a century, Britain extracted an estimated £100 million from India; and in 1585, one-quarter of Spain's total revenue came from its American colonies.
2 Land, in many cases, was acquired by force. Many indigenous people, such as the Maoris in New Zealand and the Aborigines in Australia, now find themselves owning little of their land and are increasingly pushed in to what are no more than ghettos.
3 Colonialism led to the widespread forced transplantation of people. Very large numbers of slaves were taken from Africa to other parts of the world, principally to the Americas. One estimate is that between 1601 and 1870, about 15.2 million slaves were taken from Africa. This slavery not only decimated some native populations, but also introduced population plurality in hitherto homogeneous populations. Other forced labour systems, such as the system of semi-slavery known as the indenture system, also led to the transplantation of a large number of people.

4 Colonialism deprived the populations of their right to determine their own future. These decisions were made by colonizers, often in the context of the self-interest of the metropolitan powers.

5 The colonial system was based on racism – the belief that one race was superior to others. Nearly every facet of colonial social, political and economic organization reminded the native populations of their inferior status. Examples include, in addition to the justification of lack of self-determination on racial grounds, the payment of lower wages to indigenous workers compared to white people and the refusal of entry to institutions such as schools and clubs. All these practices engendered in the populations a sense of inferiority in themselves and their own people, and a sense of confidence in European people and things.

6 Linkage with the world system enabled colonies to trade with the rest of the world, and thus to obtain commodities they were unable to produce themselves. It also enabled them to dispose of their surplus production, in addition to being able to specialize in the production of those things in which they had some comparative advantage. However, these links also led to the development of primary products economies, for agricultural trade opportunities were then used as the basis for continuing to develop these societies as agricultural societies.

7 The colonial powers introduced physical infrastructure, such as railways, roads and sea transport facilities, all designed to enable a better outward movement of primary products and the inward movement of imported goods. Despite the self-interested motivation, the infrastructure also helped the local population, especially after independence.

8 The educational and legal systems introduced by the colonial powers formed the basis of subsequent development. Admittedly these systems were designed to serve the interests of the colonial powers, but they were used by these societies even after independence.

9 The colonial powers did reduce the incidence of warfare among tribes, and established law and order which was critical to the development of these societies. Further, modern states were also created, though not without problems later.

Case study B

Colonialism and industrialization in Kenya

Colonialism

Before contact with the western world Africa consisted of small tribal groups, states and some empires; the level of development among the different parts of Africa was uneven. The British were the first European power to move into this area (Asian and Arabian traders had long been active in Africa), but they were not initially keen to establish formal political and military control over East Africa, which comprised the regions that eventually became Kenya, Uganda and Tanganyika. The British government was much more interested in pursuing settlement and trade in this area under the Sultanate of Zanzibar without the threat from other powers, hoping to achieve its commercial objectives without direct administration. However, by the late nineteenth century new European powers were actively establishing empires. France, Germany and Belgium were all interested in establishing colonies in East Africa. Germany, in fact, had obtained a firm foothold by 1884. The growing threat of the establishment of colonies by rival powers, and the greatly increased commercial interest of British settlers and traders, together with additional information on the economic potential of the interior, led the British to intensify their campaign to establish formal control over East Africa. They succeeded in doing so in 1895 with the proclamation of the East Africa Protectorate. It was in 1920, however, that Kenya was formally annexed as a separate colony.

Colonialism and Kenyan society

The immediate concern of the colonial administrators was to make the colony financially self-sufficient, to which end they proposed to develop commercial agriculture. The problem lay, however, in the acquisition of land and labour; capital was to be supplied by settlers. The government solved the land problem by declaring crown ownership of land and by giving white settlers long leases, the initial 99 year leases being converted into 999 year leases. The

Case study B *(continued)*

choicest agricultural land was thus alienated to Europeans, as had already happened in other colonies.

The government attempted to solve the labour problem by limiting the amount of land left to Africans, hoping that they would be forced to work for Europeans, and by imposing taxes on the indigenous population, which they would have to pay by working for wages. These measures were strengthened by the imposition of higher customs duties on imported goods, which attempted to raise the cost of living of Kenyans, again with the expectation that they would work for wages. The colonial government did succeed in creating wage labourers out of Africans, a process known as proletarianization.

The general pattern of development in Kenya was typically colonial. The colony produced a restricted range of agricultural crops, and a condition known as a monocultural economy. This economy was closely tied to Britain, and most of its linkages were external rather than internal. Europeans occupied the upper echelons of commerce and administration, followed by Asians, upon whom the British relied to provide skills the Africans lacked, followed by Africans at the bottom.

Manufacturing

When Africans first came into contact with Europeans, they already possessed basic manufacturing. They were able to produce a wide range of woven products; metal products, such as knives, farm implements, jewellery and chains; leather goods; and complex pottery. In other words, they produced most of their required manufactured goods. Europeans often admired their craftsmanship and sometimes employed them as well.

The British government had no interest in promoting industrialization in Kenya. British colonial policy there was conditioned very much by its interests and circumstances; what Britain needed were raw materials for her factories and protected markets for her manufactured goods, especially as she could not rely on trade with other European countries, which were trying to protect their own industries by limiting British imports; and places for investment.

Case study B *(continued)*

British colonial policy made it quite clear that it was not in its interest to encourage the industrialization of Kenya, which it hoped would remain an agricultural region.

So what happened to traditional manufacturing? There was a process of deindustrialization as the colonial government slowly killed it off both by flooding the market with cheap factory-produced goods and by controlling the trading system. Even Kenyans began to see their profits in selling imported manufactured goods in the place of locally manufactured goods.

However, some manufacturing did develop, mainly in the area of processing agricultural commodities. Thus textile and oil mills were established. In addition, local initiatives led to the establishment of some other industries, such as the making of boots and shoes, taking advantage of the availability of leather, and matches. These were established as import-substituting industries, meaning that they were replacing items previously imported. They enjoyed some natural protection from imports because of the relative distance of Kenya from British factories. Manufacturing received added impetus during the two world wars, which necessitated some industrialization due to shipping difficulties and which also gave opportunities for industrialization. However, these factories were not established as a result of British encouragement; they were established simply by local initiative seizing sound commercial opportunities.

Colonialism and industrialization

We have discussed the impact of colonialism on the Third World in general terms. In this section, we shall examine the impact of colonialism on industrialization in the Third World. The issue of the impact of colonialism on industrialization can best be raised by answering the following questions. What impact did colonialism have on existing manufacturing in the Third World? Did colonialism promote the development of manufacturing in the Third World; did it stifle its growth; or, indeed, was it neutral towards industrialization?

When Europe colonized what is today the Third World, the bulk of these areas did not have sophisticated manufacturing economies. For such countries, the issue of the impact of colonialism on existing manufacturing systems is irrelevant. However, there were countries that did have highly sophisticated manufacturing economies. India is an excellent case in point. Indian manufacturing was famous for its craftsmanship and its output enjoyed an international reputation. A large proportion of its labour force was engaged in the manufacturing sector. Britain destroyed this thriving and sophisticated manufacturing economy to induce India to import manufactured goods from her. Some idea of the impact of these policies can be obtained from the fact that the proportion of the Indian working population engaged in industry and handicrafts fell from about 25 per cent at the beginning of the nineteenth century to about 18 per cent in 1881. It is interesting that when Britain was undergoing its industrial revolution it banned the import of Indian textiles, arguing that British manufacturers could not compete with the superior Indian textiles and needed protection. However, when Britain was in a position to mass-produce textiles, it flooded the Indian and other markets without any regard to what would happen to indigenous manufacturers.

It is generally agreed that colonialism did make possible, albeit unintentionally, the transfer to the colonies of elements of the industrial economy. In the case of India, for instance, Britain promoted the development of railways with a view to exploiting Indian raw materials more intensively. Colonialism also initiated some limited degree of the industrial processing of raw materials to reduce bulk. One can also argue that many infrastructural developments, whatever the motive for their creation, provided a base on which industrial development in the post-colonial period took place.

It is now generally agreed that it was in the interest of metropolitan powers not to encourage industrialization in their colonies, not only because this would divert labour from the production of raw materials needed by them, but also because it would reduce the market for manufactured goods. Colonial raw materials and protected markets were of vital importance to metropolitan industries.

There are many clear instances, stretching over a long period, that indicate a great resolve on the part of colonial powers to prevent industrialization in other colonies besides India. In North America, for instance, an Act of 1699 forbade the export of cloth manufactures from the colonies, even to another American colony. In 1750, the erection of new slitting or rolling mills, forges or furnaces was made illegal. In another of many such moves, in 1732 the export of hats made in America was banned, as was the making of salt in India, so that industrial salt from Britain could be sold. More recently, in Kenya a Japanese company was denied permission to establish a match factory because match companies in the United Kingdom objected; similarly, a twine factory in Tanganyika was closed in 1936 when British twine manufacturers complained about it.

We can see, therefore, that colonial powers had a definite interest in ensuring that minimal industrialization took place in the colonies. The colonial powers either reduced manufacturing in colonies where it existed, or discouraged its development, such as through allocation of greater resources to agriculture and refusal to grant protection and provide other assistance. It is not surprising, therefore, that industrialization in most developing countries started in earnest only after they had attained political independence. This is not to say that there was no industrialization during the colonial period – for this is not true – but colonialism did constitute a major constraint.

Independence and industrialization

The importance of the attainment of political independence by developing countries to their attempts to industrialize cannot be overemphasized. According to the renowned economist Keith Griffin (1981): 'Without decolonization there probably would have been no development economics or development studies and possibly not much development either.' Most developing countries had adopted development planning as their route to rapid economic development from 1930 onwards following its adoption by the Soviet Union. Mindful of the neglect and discouragement of industrialization during the colonial era, and recognizing its vital role in meeting quickly the demands of their large and burgeoning populations, these countries quickly adopted ambitious industrialization plans.

The majority of developing countries adopted an import-substitution approach to industrialization, whereby those items previously imported were targeted for local production. With protection to local manufacturers ranging from high tariff rates on imports to their total ban, this proved both popular and initially successful. However, this approach also engendered an inward-looking mentality among industrialists, and the absence of competition meant innovation was stunted. This resulted in inefficient manufacturing economies with high costs and a lack of competitiveness.

In the late 1960s, and especially in the 1970s, developing countries began to review their industrial policies and found that reliance on import-substitution industrialization had many disadvantages. There appeared to be excess capacity in the manufacturing sector. The lack of competition resulted in a lack of innovation and in stagnation. Countries that had followed more open, export-oriented industrialization strategies appeared to have fared much better than those following import-substitution industrialization. On the whole, therefore, industrial policies shifted towards export-oriented industrialization. The actual progress of the efforts of these countries and the nature of Third World industrialization are discussed in the next chapter.

Key ideas

1 The historical relationships between the now developed countries and the developing countries are very important in explaining the degree and nature of contemporary Third World industrialization.
2 Most countries of the contemporary Third World were colonies of the present industrialized countries. Colonialism started in the Americas in the fifteenth century but accelerated in the nineteenth century, in the wake of Europe's industrial revolution.
3 Colonialism had a mixed impact on industrialization in developing countries, providing the physical infrastructure and socio-economic and political institutions for later industrialization, but at the same time restricting indigenous enterprise.
4 Industrialization in developing countries began in earnest only after they obtained political independence.

3
The level and structure of Third World industrialization

Introduction

No aspect of development planning in the Third World has been as widely adopted and persistent as industrialization. Nearly all governments, large and small, are fully committed to at least a degree of industrialization to meet more adequately the rapidly rising aspirations of their burgeoning populations. As we will see, while considerable progress has been made in Third World industrialization, success has graced only a few countries, and the vast majority have made hardly any progress.

The level of Third World industrialization

We can examine the level of industrialization in the Third World in two ways. The first is to examine the importance of manufacturing production in terms of GDP (see Table 3.1). Two main points emerge from this table. First, manufacturing still constitutes only a small part of the national economy for most developing countries; and, second, there has not been a major change in the importance of manufacturing in the GDP of developing countries in the last two decades, the proportion changing by only 1.9 percentage points. By way of comparison, we should note that, in 1983, manufacturing in developed market economies contributed 25 per cent to GDP. We can say, therefore, that there has been little structural change in the economies of developing countries as a whole in the last two decades.

Table 3.1 The share of the manufacturing sector in the GDP of developing countries and territories

Year	Share in the GDP (%)
1960	15.6
1965	16.9
1970	18.1
1975	18.1
1980	18.1
1983	17.5

Source: United Nations Conference on Trade and Development (1987) *Handbook of International Trade and Development Statistics, 1986 Supplement*, New York: United Nations.

Figure 3.1 Share of developing countries in world manufacturing value added and trade

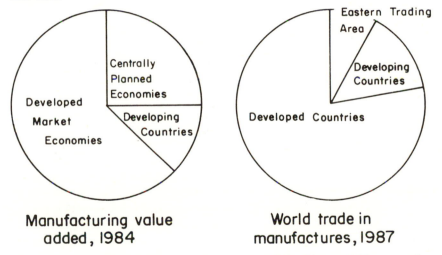

Manufacturing value
added, 1984

World trade in
manufactures, 1987

Sources: United Nations (1985) *Industry in the 1980s: Structural Change and Independence*, New York; General Agreement on Tariffs and Trade (GATT) (1988) *International Trade 1987–8*, Geneva.

The second, equally important, index of Third World industrialization is its share of global manufacturing value added and trade, presented in Figure 3.1. Developing countries still account for a very small proportion of world manufacturing production: 11.6 per cent in 1984. It is equally clear that in the last two decades there has not been

Figure 3.2 Share of developing countries in world manufacturing value added at constant (1975) prices

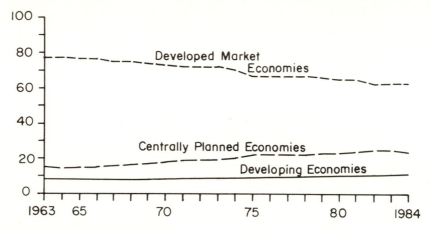

Source: United Nations (1985).

a major change in the Third World's contribution to world manufacturing value added (Figure 3.2). This is not to say that there has been no increase in industrial expansion in the Third World, for clearly there has, but that the share of Third World production of global manufacturing production has not changed significantly.

From Figure 3.3 it is clear that, although the Third World's share of global trade in manufactures remains small (14 per cent in 1986), there has been a rapid improvement in its manufacturing trading position: a change of 219 per cent between 1963 and 1987, albeit on a small base.

This difference in the manufacturing trade performance of developing countries and their share of global manufacturing value added is interesting. It points to the tendency among the developing countries towards export-oriented industrialization. The overall industrial production in developing countries is not expanding much faster than in industrialized capitalist and centrally planned economies, but these countries are exporting more of their production.

Regional contrasts in Third World industrialization

The Third World comprises a diverse group of countries with different resource endowments; international alliances and linkages; comparative

Figure 3.3 Share of developing countries in world exports of manufactures

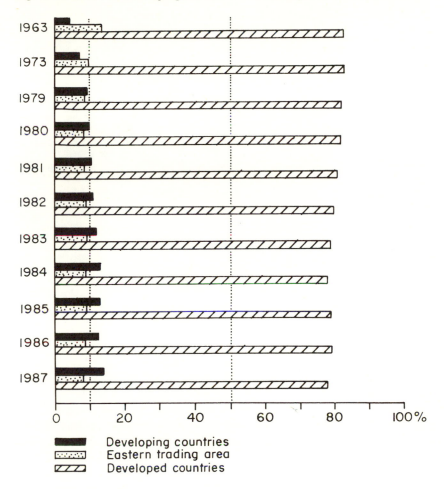

Source: GATT (1988).

advantages; and with different public policies. It is not surprising, then, that there have been marked contrasts in the performance of developing countries in industrialization. We should be aware of this spatial diversity.

Tables 3.2, 3.3 and 3.4 portray spatial diversity in manufacturing in the Third World. The very major contrast between the manufacturing

Plate 3.1 A copra factory in Majuro, Marshall Islands. Resource processing forms the bulk of manufacturing in many developing countries.

performance of middle-income countries (defined by the World Bank as countries with per capita GNP in 1986 of US$426 or more) and low-income countries is apparent. The middle-income countries have consistently performed better in manufacturing production and exports. Not only this but, unfortunately, the gap between the manufacturing performance of middle-and low-income countries has been widening. For instance, whereas in 1965 the share of world manufacturing value added of the two groups was roughly equal, in 1985 the middle-income countries' share was almost twice that of the low-income countries. Within the latter, the least developed countries have done particularly badly; their share of world manufacturing value added has remained unchanged at about 0.2 per cent in the last 20 years.

Table 3.3 brings out the regional contrasts in Third World industrialization. The African continent has fared the worst; while Latin America, which had a headstart in industrialization because of its early independence and its proximity to a major centre of production and consumption (USA) has done reasonably well. However, while some

Table 3.2 Regional contrasts in manufacturing in developing countries

Region	Share in manufacturing production			Share in manufactured exports		
	1965	1973	1985	1965	1973	1985
Developing countries	14.5	16.0	18.1	7.3	9.9	17.4
Low-income	7.5	7.0	6.9	2.3	1.8	2.1
Middle-income	7.0	9.0	11.2	5.0	8.1	15.3

Source: World Bank (1987).

Table 3.3 Regional share of world manufacturing value added

Region	Share of manufacturing value added		
	1975	1980	1985
Caribbean and Latin America	5.70	6.00	5.37
Tropical Africa	0.44	0.42	0.40
North Africa and West Asia	1.29	1.20	1.58
Indian Subcontinent	1.23	1.13	1.27
East and South-East Asia	1.67	2.43	3.26
Total developing countries	10.33	11.18	11.88

Source: United Nations Industrial Development Organization (1988) *Industry and Development: Global Report, 1988/89*, Vienna: United Nations Industrial Development Organization.

Table 3.4 The leading developing country exporters of manufactures in 1986

Country	Value of exports in 1986 (US $ million)	Percentage of total developing country manufactured exports
Republic of Korea	22,996	22.3
Hong Kong	17,181	16.7
Mexico	9,904	9.6
Brazil	9,290	9.0
Singapore	7,293	7.1
Yugoslavia	5,173	5.0
Malaysia	4,044	3.9
Thailand	3,027	2.9
India	2,925	2.8
Philippines	2,773	2.7
Indonesia	2,002	1.9
Pakistan	1,477	1.4
Argentina	1,155	1.1
Total developing countries	103,033	

Source: United Nations Conference on Trade and Development (UNCTAD) (1988) *Protectionism and Structural Adjustment; Problems of Protectionism and Adjustment*, New York: UNCTAD.

Plate 3.2 A perfume factory in the tiny Cook Islands in the South Pacific. Most of the output is sold to tourists.

countries in Latin America have performed very well, such as Brazil, Mexico and Argentina, the most dynamic growth in industrialization in the last decade has come from South-East Asia, or, more specifically, from the four tigers: Hong Kong, Singapore, Taiwan, and South Korea (see Case study C).

Case study C

South Korean industrialization

South Korea has an area of just under 100,000 km^2 and in 1987 had a population of about 42 million people. It is, therefore, not a large country either in size or in population. However, South Korea, a leading NIC, has achieved spectacular industrialization since the 1960s and is looked upon by other developing countries as an inspiration in their struggle for industrialization. The South Korean case is frequently used to show that it *is* possible for developing countries to break the vicious cycle of poverty, to

Case study C *(continued)*

achieve structural change and to narrow the economic gap with the developed countries.

Like most other developing countries, South Korea began its drive for industrialization with import substitution, which resulted in a rapid growth in production as entrepreneurs moved to take advantage of very attractive conditions with low risks. Also, given the high levels of protection, there was no problem with overseas competition. Many of the goods produced did not need economies of scale or a skilled labour force. As in most other countries, the early opportunities in producing low-level goods were quickly exhausted. To proceed, Korea needed to produce higher-level goods, such as durable goods like refrigerators, and intermediate goods which were to be used in further manufacturing. These were problematic because they needed larger markets and called for technical sophistication.

Plate 3.3 A major textile factory in South Korea. Textiles have formed a major part of South Korea's industrial success.

Case study C *(continued)*

By the late 1950s, South Korea had exhausted its easy phase of import-substitution industrialization. Unlike many other developing countries that proceeded to the less advantageous second stage of import-substitution industrialization, South Korea decided instead to pursue an export-oriented strategy. This was done through an overhaul of the system of incentives, leading to a reduction of protection for domestic industries. The currency was also devalued, making Korean products cheaper and therefore more competitive in overseas markets. The devaluation also helped the agricultural sector since farmers received higher prices in local currency for their export products. Overall, then, the new package of incentives made it more attractive to export manufactured goods than to produce for the domestic market.

The Korean success in industrialization can be seen both in the

Table 3.5 The growth in Korean manufacturing, 1963–84

Year	No. of establishments	No. of persons engaged (000)	Gross output (billion won)	Value added (billion won)
1963	18,310	402.2	167.0	61.4
1967	23,455	643.7	548.0	206.6
1968	23,808	741.6	767.3	300.1
1969	24,752	820.1	1.045.1	424.2
1970	23,905	854.1	1,331.6	547.9
1971	23,208	824.4	1,669.9	688.6
1972	23,729	973.7	2,241.5	899.3
1973	22,293	1,157.8	3,697.0	1,397.6
1974	22,632	1,298.4	5,704.0	1,867.2
1975	22,787	1,420.1	8,165.0	2,828.1
1976	24,957	1,717.3	11,679.0	4,075.1
1977	26,726	1,918.6	15,438.0	5,596.9
1978	29,864	2,111.9	21,157.0	8,193.0
1979	31,804	2,116.7	26,692.0	9,205.0
1980	30,823	2,014.7	36,279.0	11,587.0
1981	33,471	2,044.2	46,716.0	15,412.0
1982	36,799	2,099.1	51,651.0	17,306.0
1983	39,243	2,215.9	60,547.0	20,912.0
1984	41,549	2,343.4	71,308.0	24,656.0

Source: United Nations Department of Economic and Social Affairs (various years) *The Growth of World Industry* and *Yearbook of Industrial Statistics*.

Case study C *(continued)*

rapidly increasing importance of manufacturing in the national economy and in exports. Table 3.5 shows the rapid expansion of the manufacturing sector, with the manufacturing value added increasing more than 340 times in 20 years. Figure 3.4 demonstrates a significant structural change in the economy: the manufacturing sector's contribution increased from 19.5 per cent in 1973 to 30.4 per cent in 1984.

The export performance of South Korea has been similarly outstanding. Between 1960 and 1969, it increased its manufactured

Figure 3.4 South Korea: manufacturing share of GDP

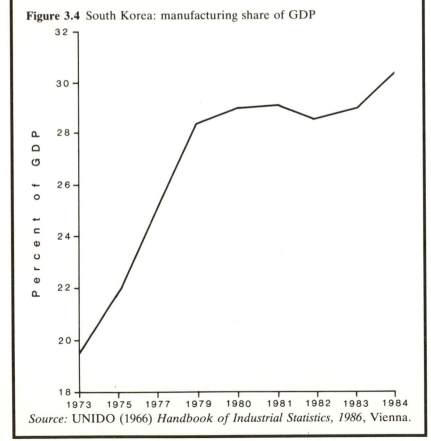

Source: UNIDO (1966) *Handbook of Industrial Statistics, 1986*, Vienna.

Case study C *(continued)*

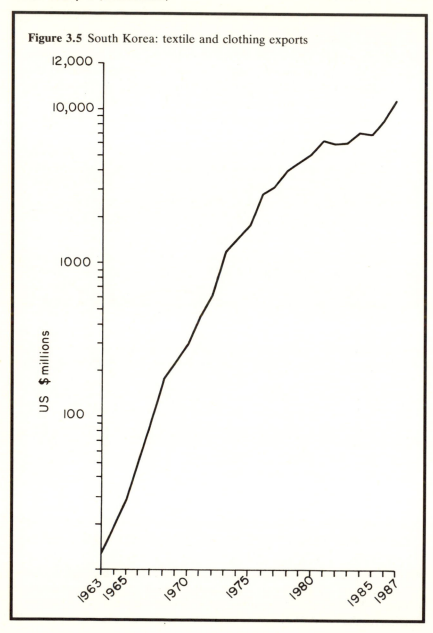

Figure 3.5 South Korea: textile and clothing exports

Case study C *(continued)*

Plate 3.4 A shoe factory in South Korea, an example of a labour-intensive industry.

exports at an annual rate of 69 per cent compared with the rate of developing countries as a whole of 8 per cent per annum. The Korean success in exporting can be seen in its textiles and clothing exports, in which it had a comparative advantage because of its cheap labour (Figure 3.5).

So phenomenal has its export success been that it has risen from being the world's twenty-sixth largest exporter of manufactured goods in 1970 to the twelfth largest in 1987. Moreover, South Korea has diversified its exports and has become the world's sixth largest exporter of electronics, a dramatic and very significant achievement (Figure 3.6). It has also become the world's second largest builder of ships, and is emerging as one of the developing world's major automotive exporters. It is now undergoing a vital transition from its past as a cheap producer of goods to a mature industrial economy. It is interesting to note that it has not only

Case study C *(continued)*

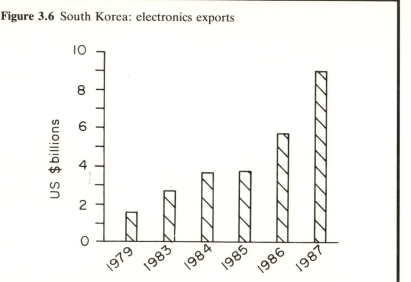

Figure 3.6 South Korea: electronics exports

changed economically, but also undergone major political transformation in the last few years, moving from a dictatorship to a
democracy, reflecting its changing position as a world economic
power.

The case of South Korea provides considerable inspiration to
other developing countries, although there were peculiar circumstances in this case that are not present for other developing
countries, such as an early start in exporting and massive assistance from the United States of America because of South Korea's
strategic importance. However, government policy, especially in
the use of exchange rates and the timely overhaul of the incentive
system, can serve as a model for other developing countries.

Some idea of the uneven distribution of manufacturing in the Third
World is given by the fact that, in an analysis of the key Third World
industrializing countries in 1981, it was found that only one country,
Brazil, accounted for 21 per cent of the total manufacturing value added
of the seventy-three developing countries, and it, together with Mexico,
India, Argentina and South Korea, accounted for 58 per cent of the

Figure 3.7 Manufacturing production of over US $5 billion in developing countries, 1983

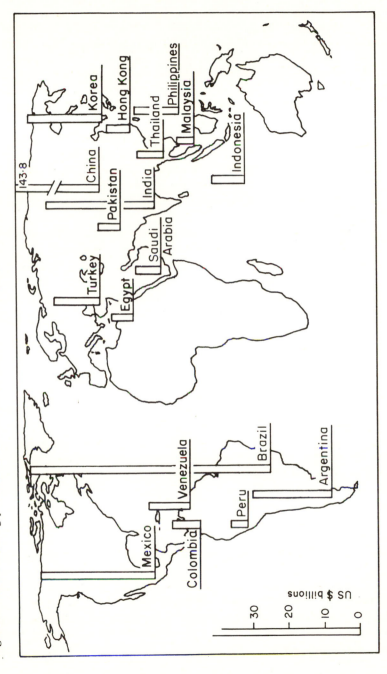

Source: UNIDO (1986) *Handbook of Industrial Statistics*, Vienna.

combined manufacturing value added of these seventy-three countries. Furthermore, thirteen developing countries (excluding China because data on it are not available) identified as major industrial developing countries contained an incredible 78 per cent of the total manufacturing value added of the seventy-three developing countries mentioned. Figure 3.7 shows the main developing countries in terms of manufacturing value added.

The very uneven nature of Third World industrialization is also reflected in exports, where a small group of developing countries account for most of the exports from developing countries, as shown in Table 3.4. The same concentration can be seen here; three countries account for nearly half of all exports, while thirteen account for 86 per cent of all exports.

This uneven development of manufacturing in the Third World is one of its most striking features. It explains the contradictory assessment of Third World industrialization: authors who look at general Third World development generally argue that there has been little industrialization, while those who focus on the newly industrializing countries (NICs) often exaggerate the degree of industrialization.

Reasons for differential growth

After examining the major contrasts in industrial performance among developing countries, it is natural to ask why this should be so. Certainly, the countries that are not performing well are keen to identify the ingredients of a successful industrial strategy. The reasons for the success of the small number of countries that we have identified are difficult to find; however, a number have been put forward.

Early independence has been important in explaining some of the success of industrialization; after all, colonial governments had no interest in the industrialization of colonies. Latin America's high level of industrialization is attributed to this early independence and its consequent head start.

The *size of the domestic market* has been and remains an important factor in Third World industrialization. The poorer countries do not have a large market, because of their small populations and their low per capita incomes. This small domestic market has prevented the development of industries that require large production runs; that is, there are no economies of scale. Consequently, there has been both duplication of industries and monopolies. This is now, however, not seen as a major impediment to industrialization because of the possibility of overcoming the problems of small scale by exporting.

Most *foreign direct investment* has flowed to the middle-income developing countries. For instance, only twenty developing countries account for almost 90 per cent of foreign direct investment. Most of this is to oil- or mineral-rich countries, those with large domestic markets, or those with export-oriented industrialization. Only three countries, Brazil, Mexico and Indonesia, account for about one-third of all foreign investment. Foreign investment helps industrialization by increasing manufacturing output and by providing export markets and access to new technology.

It has been noted that the *special relationship* of Taiwan and South Korea with the United States has been important in explaining some of the industrial success of these countries, because this allowed for more US investment and preferential access to the US domestic market.

Nearly everywhere, the industrial revolution has been based on sound *educational development*. This is especially important for the Third World because of its technology imports; bureaucrats and enterprise managers need to be educated to be able to make intelligent choices in what are decisions with far-reaching consequences. The NICs, in particular, have invested greatly in education over a long period of time.

Appropriate government policies have been identified as being the most crucial determinants of industrial performance in recent times – even more important than resource endowment. A marked feature of successful industrialization policies has been the clear vision and bold commitment shown by some governments. In the case of South Korea, there is a clear national consensus that they can and should do better than other Asian nations; they feel that they can overtake Japan as the major industrial power in the region. Similarly, Singapore's industrial and economic success is based to a large extent on the clearly expounded philosophy of her long-serving prime minister.

Successful industrial policies call for stable, determined, and clearly defined and consistently interpreted government policies; a sound relationship between the government, the business sector and labour; and for a balance between export orientation and import substitution. It also calls for competition in the manufacturing sector to increase its efficiency and internal competitiveness.

The structure of Third World manufacturing production

The first aspect that most industrial geographers examine is the structure of the manufacturing sector. This relates to what is produced: in other words, the activities that make up manufacturing. The second aspect implied by manufacturing structure is the nature of the market: that is, the kind of market the firms operate in. The structure of the manufacturing sector is illustrated by the categories of manufacturing activity using the International Standard Industrial Classification of All Economic Activities, which we have already discussed in Chapter 1.

What are the main types of manufacturing in the Third World? Food manufacturing, which comprises both food processing, such as sugar milling and flour milling, and further recombination of food products, such as ice cream and confectionery manufacturing, is clearly the overwhelming important activity in the Third World, followed by textiles (see Figure 3.8). There are other important types of manufacturing, such as iron and steel, electrical machinery and transport equipment, but none of these stands out as being relatively more important than the others.

Why does the Third World have this pattern of manufacturing? The explanation lies in the law of comparative advantage, which stipulates

Figure 3.8 Main components of Third World industrial production, 1984

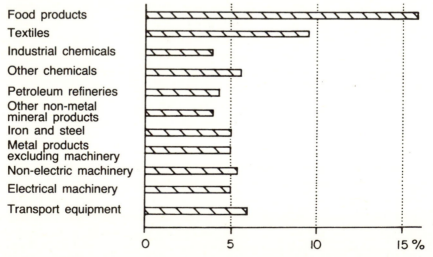

Source: UNCTAD (1988) *Protectionism and Structural Adjustment*, New York.

that a country will specialize in the production of those goods in which its *relative* advantage is the greatest. A country would normally achieve this by producing those goods which use its abundant resources the most.

Although international trade does not strictly follow the law of comparative advantage, there is merit in this proposition. The two largest categories of manufacturing in developing countries, food and textiles, indicate its relative advantage in labour-intensive goods because of its cheap labour. The importance of food manufacturing also derives from the agricultural nature of Third World societies. Wage costs in developing countries are a fraction of those in developed countries, even if adjusted for different labour productivities. This explains the predominance of labour-intensive manufacturing in the Third World, such as garments, light electronics and standardized machinery.

Industrial structure, like other segments of society, changes considerably over time. This slow shift in the industrial structure of developing countries is well illustrated in Tables 3.6 and 3.7. What changes can we detect, and why may this be happening? The main changes we can observe are that traditional manufacturing (food processing, textiles,

Table 3.6 The structure of manufacturing of developing countries for selected years

Manufacturing group	1963	1970	1978
Food products	26.3	21.2	18.9
Beverages and tobacco	12.0	9.8	10.1
Textiles	11.3	11.8	10.8
Wearing apparel	3.6	3.9	4.1
Wood and cork products	3.7	3.2	2.8
Furniture and fixtures excluding metal	2.1	1.8	1.3
Paper	1.8	2.0	2.3
Printing and publishing	2.9	2.6	2.4
Industrial chemicals	2.1	2.8	3.2
Other chemicals	4.2	4.6	5.3
Petroleum refineries	5.6	7.2	7.1
Other non-metallic mineral products	3.5	3.9	4.0
Iron and steel	1.6	2.5	2.6
Metal products excluding machinery	3.5	4.2	4.3
Non-electrical machinery	1.6	1.9	2.3
Electrical machinery	1.5	2.6	3.7
Transport equipment	2.7	3.1	3.8

Source: UNIDO (1982) *Handbook of Industrial Statistics*, New York: United Nations.
Note: Only industries contributing 2 per cent or more to total manufacturing value added are included.

Table 3.7 The structure of manufactured exports from developing countries

Export category	Share of developing countries' exports (%)		Growth rate (%)
	1970	1980	1970–80
Traditional manufactured exports:			
Labour-intensive:			
Textiles and apparel	31.3	24.8	11.8
Footwear	1.8	2.9	18.2
Other labour-intensive	2.9	2.3	11.6
Total	36.0	30.0	12.4
Resource-based:			
Wood and cork	3.6	1.5	6.9
Paper manufacture	0.8	1.1	17.6
Other resource-based	0.8	0.9	14.5
Total	5.2	3.5	12.2
Non-traditional manufactures:			
Electrical machinery	16.1	16.7	14.1
Chemicals	8.3	9.9	15.3
Non-electrical machinery	4.2	8.7	20.1
Transport equipment	2.6	5.2	20.0
Iron and steel	6.2	6.5	14.2
Other non-traditional	21.4	19.5	12.9
Total	58.8	66.5	15.1
Total	100.0	100.0	14.0

Source: World Bank (1987).

leather products and so on) has been shrinking relative to other types of manufacturing, which have been increasing. For instance, chemical products have become important, as has electrical machinery.

These changes are part of the structural changes in a society as it develops. Agriculture, for instance, has become a smaller part of the economy worldwide as other non-agricultural activities are developed to diversify the economy. In part, they also reflect the growing Third World comparative advantage in some skill-intensive types of manufacturing such as mid-range electronic items and capital goods.

Heavy manufacturing or capital goods manufacturing has become important, particularly in countries such as South Korea, Brazil, China and India. Some indication of the general trend towards heavy industry

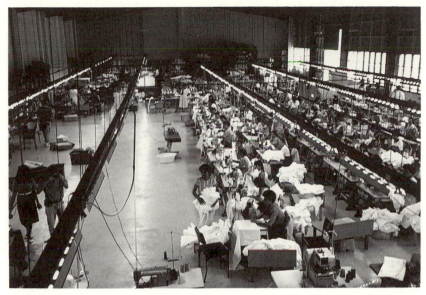

Plate 3.5 Garment manufacturing forms an important part of Third World manufacturing. This plate shows a garment factory in the small island country of Fiji, which has dramatically increased its garment exports in the last four years.

in developing countries is given by the fact that the share of heavy manufacturing rose from 33 per cent in 1955 to 43 per cent in 1963 and reached 57 per cent of manufacturing value added in 1978.

So far we have dealt with the issue of the structure of manufacturing in the Third World in terms of what is produced, that is in terms of the structure of production. Another element of the structure of manufacturing is the structure of producers, the market structure. In what type of markets do manufacturers produce in the Third World? Information on the manufacturing market structure of developing countries is generally not available but Case study D gives a brief report on how this operates in Fiji, a rapidly expanding centre of production.

Case study D

Manufacturing market structure in Fiji

Fiji is a small island nation in the South Pacific with a population of about 750,000 people and a land area of just over 18,000 km^2. For the manufacturing sector as a whole, one government-owned firm, Fiji Sugar Corporation, accounts for an overwhelming share of manufacturing value added and employment. In 1984, it contributed 25 per cent of the national manufacturing value added. If we were to add to this the five other major manufacturing

Table 3.8 Fiji: the degree of concentration in manufacturing, 1981

Industry	Level of concentration
Edible and coconut oil	Very high
Sugar milling	Very high
Beer, cigarettes, tobacco	Very high
Paint	Very high
Repairs and maintenance of industrial machinery	Very high
Butchering and meat packing	High
Rice and flour milling	High
Footwear	High
Paper products	High
Retreading and flip-flops	High
Shipbuilding and repairs	High
Bus building	High
Dairy, fruit and fish	Fairly high
Bakery products	Fairly high
Confectionery	Fairly high
Miscellaneous food	Fairly high
Non-alcoholic drinks	Fairly high
Soap, toiletries and chemical products	Fairly high
Plastics	Fairly high
Cement and concrete products and basic metals	Fairly high
Metal furniture and fixtures and structural metal products	Fairly high
Curios and artifacts	Fairly high
Agricultural machinery and equipment	Fairly high
Miscellaneous products	Fairly high
Sawmilling	Fairly low
Printing and publishing	Fairly low
Fabricated metal products except machinery and equipment	Fairly low
Furniture and uphostery	Low
Textiles and clothes	Low

Source: Chandra, Rajesh (1988) 'Industrialisation in Fiji: industrial structure and organisation', *GeoJournal* 16:2.

Case study D *(continued)*

corporations in Fiji, probably more than three-quarters of the total manufacturing value added originates from these corporations. We can say, then, that there is a high degree of concentration in the manufacturing sector, meaning that a small number of producers dominate it.

At the individual industry level, which is a better level to examine market structure, there is also a very high level of concentration. Of the thirty industry groups on which census data are reported, only two have low levels of concentration, namely furniture and upholstery, and textiles and clothes (Table 3.8). These are, of course, industries where only small amounts of capital are needed to start production. Apart from these sectors, a high level of concentration exists in the manufacturing sector.

Not only is the industrial structure highly concentrated, but also we can characterize the entire manufacturing sector in Fiji as a combination of monopolies and oligopolies. In the case of monopolies, the market is controlled by a single producer or seller, while oligopoly refers to a situation where a small number of producers control the market for a particular good. We can see the level of monopolies and oligopolies most clearly at the level of products. Of the ninety-six products identified in a survey of manufacturing in Fiji in 1983, 27 per cent were produced by firms enjoying a monopoly in production and another 30 per cent had oligopolistic producers. Thus more than half of the products identified had either monopolistic or oligopolistic market structure.

The high degree of concentration is to be expected in a small market where producers are heavily dependent on the domestic market. Economies of scale demand plants of certain size and this precludes the existence of numerous plants. In many instances, the government guarantees protection to firms by prohibiting the establishment of similar enterprises for a specified period of time.

Key ideas

1 While there has been significant industrial expansion in developing countries, their share of world production and trade has not improved very much since the 1960s.

2 There is marked regional variation in Third World industrialization, with the NICs monopolizing most of the manufacturing production and trade.

3 While most of the Third World's present manufacturing still reflects its traditional comparative advantage in labour-intensive production, there is a change in the range of its manufacturing, which now includes some areas of complex technology.

4
The organization of manufacturing in the Third World

Introduction

The previous chapter examined the level and structure of Third World industrialization. This chapter describes the organization of manufacturing in the Third World in terms of the ownership of manufacturing enterprises, focusing on the role of foreign investment and the state ownership of manufacturing activities. The chapter will then examine the forms of commercial organization in the manufacturing sector, with particular reference to informal and small-scale manufacturing.

Foreign investment

Although Third World industrialization has been based mainly on domestic resources, the shortage of capital, technology and international marketing expertise, and the urgency of the development question have forced most developing countries to seek external financing and other assistance for their development.

Foreign direct private investment (FDI) is the major component of private capital inflows to developing countries, although FDI as a proportion of total financial inflows (which includes grants and concessional loans) has been decreasing in recent years. None the less, FDI in developing countries grew rapidly in the 1970s and is vital if they are to come even close to meeting the targets set by Lima Declaration, that at least 25 per cent of the world's manufacturing value added should be

Figure 4.1 Foreign direct investment inflow by major regions

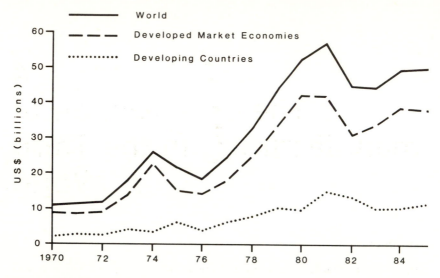

Source: United Nations Centre on Transnational Corporations (1988)
Transnational Corporations in World Development, New York: UNCTC.

located in developing countries by the close of the twentieth century.
However, the flow of FDI to the Third World decreased in the 1980s
(Figure 4.1). The Third World's share of global FDI has remained static
at about one-third since the 1960s.

Multinational corporations are extremely large organizations. The 56
largest MNCs have sales ranging from US $10 billion to US $100 billion.
MNCs are extremely important in world production and trade. It is
estimated, for instance, that the largest 600 industrial corporations
account for between 20 and 25 per cent of the production of goods in
the world market economies. They are even more dominant in international trade; it has been estimated that 80–90 per cent of exports from
the United States and Britain are associated with MNCs. It is easy to
see, then, why MNCs are of such concern and interest to the Third
World.

Where do the MNCs originate? Most MNCs come from developed
market economies (Table 4.1) although Case study E (pp. 59–62) shows
that Third World multinational companies are also emerging. MNCs
come overwhelmingly from the United States of America, which

accounted for almost one-half of world FDI flows in 1985. In more recent years, Japan's position as an investor in developing countries has improved as it tried to overcome the problem of increasing labour costs in Japan, and to take advantage of market access provided by developing countries to developed country markets under trade and aid agreements. Between 1982 and 1986, Japan's yearly foreign investment flow jumped from US $7.703 billion to US $22.320 billion, almost a three-fold increase.

Table 4.1 Foreign direct investment flows by MNC origin, 1985

Source country	Investment flow in 1985 (US $ million)	Share of the total (%)
United States	15,236.1	25.4
United Kingdom	11,206.3	18.7
Japan	6,427.1	10.7
Federal Republic of Germany	4,934.6	8.2
Canada	4,913.2	8.2
Switzerland	3,627.8	6.1
Netherlands	3,175.0	5.3
France	2,201.3	3.7
Italy	1,833.7	3.1
Australia	1,824.6	3.0
Sweden	1,278.3	2.1
Norway	1,034.6	1.7
Other countries	2,196.6	3.7
Total	59,889.2	99.9

Source: United Nations Centre on Transnational Corporations (UNCTC) (1979) *Transnational Corporations in World Development; Trends and Prospects*, New York: UNCTC.

The parentage of MNCs is important because in addition to relying on their governments to fight their battles in international affairs, such as in seeking protection from expropriation, market access and even political conditions conducive to their increased operations and profitability, they themselves can interfere in the running of a government. For instance, it has been reported that MNCs active in Brazil are spending millions of dollars to campaign against some provisions of the Brazilian draft constitution which might adversely affect their interests.

Foreign direct investment flows only to a handful of developing countries, and the degree of concentration appears to have increased over time. According to the United Nations Centre on Transnational Corporations, only twenty states account for almost 90 per cent of all

Figure 4.2 Regional flows of gross foreign direct investment to developing countries, 1983

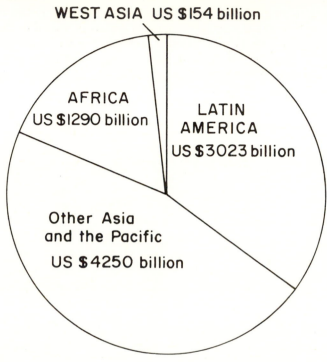

WEST ASIA US $154 billion

AFRICA US $1290 billion

LATIN AMERICA US $3023 billion

Other Asia and the Pacific US $4250 billion

Source: UNCTC (1985) *Trends and Issues in FDI and Related Flows*, New York.

FDI flows to developing countries. In general we can say that FDI is attracted to resource-rich countries, such as oil-producing countries; to countries with large domestic markets, such as Brazil and China; to countries with export-oriented industrialization, such as Hong Kong, Singapore, South Korea and Taiwan; or to generally middle-income countries with well-developed infrastructure.

Figure 4.2 shows the geographical breakdown of FDI to developing countries in 1983, while Table 4.2 provides more details on it. Together they reinforce the impression of the highly concentrated nature of FDI, as only three countries (Brazil, Malaysia and Singapore) accounted for 43 per cent of all FDI in developing countries and territories in 1983 (of US $10.04 billion; with Mexico, Egypt and Hong Kong, this rises to 63 per cent), although the presence of MNCs in the manufacturing

Table 4.2 The inflow of gross direct foreign investment to developing countries, 1983

Country	Gross inflow (US $ million)	
Latin America		
Argentina	184	
Brazil	1,557	
Chile	152	
Colombia	390	
Mexico	490	
Trinidad and Tobago	250	
		3,023
Africa		
Angola	117	
Egypt	864	
Gabon	112	
Nigeria	354	
Tunisia	184	
United Republic of Cameroon	151	
Zaire	138	
		1,920
West Asia		
Oman	154	
		154
Rest of Asia and the Pacific		
Hong Kong	609	
Indonesia	289	
Malaysia	1,371	
Papua New Guinea	139	
Philippines	104	
Singapore	1,389	
Thailand	349	
		4,250
Total		9,347

Source: UNCTC (1985) *Trends and Issues in Foreign Direct Investment and Related Flows*, New York: UNCTC.

Note: Countries with direct foreign investment gross inflow in 1983 of less than $100 million have been excluded.

industries of developing countries is highly variable but significant. The distribution of FDI in manufacturing in selected developing countries is presented in Table 4.3 – which reinforces our observation concerning the variability of FDI in the manufacturing industries of developing countries.

Into which types of manufacturing does FDI go? Most of the FDI used to go in mining, particularly in the petroleum sector. This has now

Table 4.3 The share of MNCs in exports of manufactures of selected developing countries (various years)

Country	MNC(%)	Year
Argentina	>30	1969
Brazil	43	1969
Colombia	>30	1970
Hong Kong	10	1972
India	5	1970
Republic of Korea	27	1978
Mexico	34	1974

Source: UNCTC (1983) *Transnational Corporations in Third World Development; Third Survey,* New York; UNCTC.

Figure 4.3 Types of manufacturing activities of US MNCs in developing countries, 1986

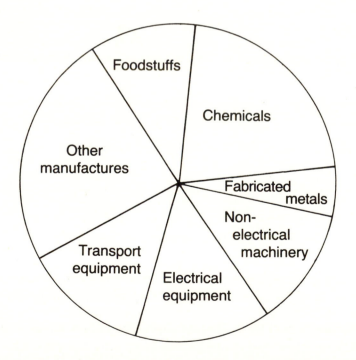

Source: UNIDO (1988) *Industry and Development Global Report,* Vienna.

changed. Of the total FDI, the share going to manufacturing varies from one source country to another, but in the case of the United States, 55.9 per cent of all new flows went into manufacturing in 1983; in the case of West Germany, the figure was 67.9 per cent; while for Japan the figure was 31.2 per cent. Thus we can say that more than one-half of the new flows are currently going into the manufacturing sector. Within manufacturing, most MNCs are active in chemicals, electrical and transport equipment, and in low labour cost items, such as garments and electronics (Figure 4.3).

There is still much debate on the role of FDI in Third World development, but it has become clear that it has a significant role to play in the context of effective government policies. Many of the concerns about the dangers of MNCs have been partially resolved by the efforts of the UNCTC, which has been instrumental in drawing up an international code of conduct for MNCs, and partly by the growing ability of developing countries to recognize, monitor and deal with these issues. The reality of Third World development and international financing is, however, that developing countries will have to draw on MNCs for capital, management and international marketing expertise for the foreseeable future. There are already indications that they are actively pursuing FDI.

Case study E

Multinationals from developing countries

The Third World has traditionally been the recipient of FDI. However, in recent years, some of the more industrialized developing countries have themselves begun to invest abroad. In 1985, for instance, developing country multinationals invested US $1.18 billion abroad, a figure that has increased rapidly from US $305 million in 1975. Many of the Third World multinationals are quite large; in 1985 there were seventeen MNCs with sales of US $1 billion or more.

The main regions investing abroad are those that receive the most FDI and have successfully industrialized. The Asian region is by far the largest investor abroad among developing countries (Figure 4.4). As would be expected, Africa has the lowest overseas investment. Only three countries (South Korea, Brazil and Mexico)

Case study E *(continued)*

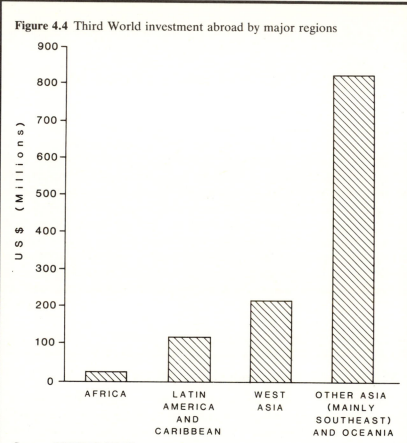

Figure 4.4 Third World investment abroad by major regions

Source: UNCTC (1988).

owned ten of the seventeen large Third World MNCs. Of these, South Korea owned six and Brazil and Mexico two each. This pattern of overseas investment by developing countries reinforces our argument in Chapter 3 regarding the highly uneven nature of Third World industrialization.

Third World MNCs invest abroad for many of the same reasons as developed country MNCs. NICs are shifting labour-intensive manufacturing to low-cost countries as their wages rise. Hong

Case study E *(continued)*

Figure 4.5 An advertisement by a major Third World multinational showing its wide-ranging technological capability

How Do You See Hyundai?

You probably picture Hyundai as a maker of affordable, high-quality cars.

That picture's not wrong. Just incomplete.

Those who work with advanced computers, ships and nuclear power plants have a better idea of our scope.

In fact, you'll find the Hyundai name behind sophisticated engineering projects, petrochemicals, robotics, and satellite communication systems, among other exciting and diverse industries.

Which is why, in Korea, Hyundai has become a symbol of our nation's economic progress.

So by all means, remember our cars. But don't forget the big picture.

HYUNDAI

K.P.O. Box 92 Seoul. Korea
TLX: K23111/5   FAX: (02) 743-8963

Source: Far Eastern Economic Review, 6 July 1989.

Case study E *(continued)*

Kong, for instance, is one of the largest investors in China. South Korea has also invested extensively in neighbouring Asian countries. Sometimes Third World MNCs try to overcome quota restrictions in developed countries by manufacturing in other developing countries not facing such restrictions, especially in textiles and clothing. Third World MNCs also try to achieve vertical integration by investing in developed countries. For instance, Petroleos de Venezuela owns 50 per cent of the stock of an oil company in West Germany as well as two oil companies in the United States.

Some Third World MNCs are investing all over the world in order to be competitive and to improve their technological capabilities. According to UNCTC, Samsung, a leading South Korean electronics corporation, had thirty-five offices in twenty-nine countries by 1980, and had begun manufacturing operations in the United States in 1984. It plans to open fifteen additional factories overseas by the end of 1990. Similarly, South Korea's Hyundai corporation has been manufacturing cars at its plant in Canada since the early 1980s (see Figure 4.5).

State ownership of manufacturing

One major aspect of industrial organization refers to the question of public participation in manufacturing, a topic that will be covered later in more detail. Here we can make some preliminary observations. In developing countries there is a high degree of state involvement, not only in industrialization but in the national economy as a whole, largely because of the lack of national middle classes capable of leading industrial development. This has often led to a proliferation of state-owned enterprises engaged in industrial production.

The importance of the public sector in the industrial economy of developing countries is indicated by their contribution to output, employment and investment. In a survey of eleven developing countries, UNIDO reported that the public sector's contribution to output ranged from 8 per cent (Mexico) to 85 per cent (Somalia), with an average of 44 per cent. The share of the public sector in manufacturing investment in a sample of twenty countries varied from 9 per cent (Morocco) to a high of 90 per cent (Egypt), with a mean of 48 per cent.

As the state's direct involvement in the industrial sector has grown, concern for the efficiency of public enterprises has emerged. In most developing countries, given the recency of the phenomenon, organization and coordination in the public sector are poor. Some countries, such as Argentina, Guyana, Malaysia and Mexico, have established separate ministries of public enterprises, while others, such as India, have established advisory and supervisory entities to make their state-owned enterprises more efficient.

Plate 4.1 Sophisticated research and development in China using high-power lasers.

There is also considerable pressure within the Third World for privatization, which involves not only increasingly private provision of previously publicly provided goods and services, but the use of private management in and the outright sale of public corporations. These pressures have grown more severe because of the alarming degree of indebtedness of the Third World, and the insistence of international financial agencies that developing countries carry out a programme of privatization as a precondition for international financial assistance.

The organization and scale of firms

The organization of commercial activities in developing countries is markedly different from the prevailing forms of economic organization in developed countries, although 'modern' forms of the latter are rapidly eroding the traditional forms. This difference between modern and traditional forms of organization has been noted for a long time, and scholars have attempted to conceptualize it in different ways. Of all the terms used, 'formal' and 'informal' sectors are the most popular.

Manufacturing activities in developing countries can be divided into non-registered, non-licensed manufacturing (the informal sector); registered and licensed sole proprietorships and partnerships (the formal sector); limited liability private companies, which display the characteristics of both informal and formal sectors; and publicly owned corporations of the kind we find in developed countries, which are definitely formal sector organizations. Despite its prominence, the modern corporation is a minor part of the economic landscape of developing countries, although it is becoming more common now with the establishment of modern stock exchanges in most developing countries. However, there is little dispute that the small-scale manufacturing subsector in developing countries is still a major component of the industrial structure in terms of both output and employment. In a study of fifteen developing countries, UNIDO has reported that the non-modern industrial sector contributed from 10 to 44 per cent of manufacturing value added. In some countries, its importance can be very great indeed. In a study by the Indian government, for instance, it is reported that 78 per cent of the Indian industrial labour force was engaged in the non-modern sector. Table 4.4 gives an indication of the important role played by small and informal sector enterprises in the manufacturing sectors of developing countries.

There is a great deal of interest in the informal sector not only because of its enormous scope for increasing industrial production, but also because of the characteristics of informal manufacturing enterprises. Informal enterprises are often locally owned, thus promoting the goal of national ownership of economic resources. They are also labour intensive, reducing the major unemployment problem and conserving foreign exchange. They also provide a training ground for local

Table 4.4 Share of small enterprises in total employment in manufacturing in selected countries

Country	Year	Percentage distribution of employment in manufacturing		
		Households and small workshops	Establishments with <100 workers	Total share
Colombia	1973	48	22	70
Ghana	1970	78	7	85
India	1973	60	18	78
Indonesia	1975	76	12	88
Kenya	1969	49	10	59
Nigeria	1972	59	15	74
Philippines	1975	53	21	74
Tanzania	1967	55	8	63
Turkey	1970	32	36	68

Source: International Labour Office (ILO) (1986) *The Promotion of Small and Medium-Sized Enterprises*, Geneva: ILO.

entrepreneurs. There are many examples of informal sector enterprises evolving into large, formal sector enterprises, although most people would argue that not enough of this happens. Finally, informal sector enterprises promote the goal of regional development given that they are present and even dominant in the smaller centres and poorer regions.

For these reasons, although most of the industrial investment in developing countries has gone into the development of the large-scale modern manufacturing sector, the majority of these countries have positive programmes to help both the non-modern subsector and the small-scale modern manufacturing units (see Case study F). Assistance has typically taken the form of industrial extension, soft financing, provision of industrial estates, often with industrial buildings already constructed for lease by this subsector, and some protection from the modern sector.

Most developing countries, however, given their development orientation, are planning to absorb their non-modern and small-scale industrial subsectors into the modern industrial sector, and to maintain only those units where economies of scale are not critical and where social justice gains are substantial.

Having said this, however, the advent of new technologies, especially the adoption of electronics and computer-aided manufacturing, which enable flexible use of machines, means that large runs are no longer necessary for efficient production. This enables small firms to become

Plate 4.2 China's first synchrotron radiation laboratory, an example of the sophisticated industrial and scientific capacity of some developing countries.

competitive with large firms and will change the way people view small-scale and large-scale manufacturers.

These developments have given small firms in developed countries a new lease of life, and developing countries are also rapidly adopting computer-aided manufacturing, making their small enterprises more competitive. How far the Third World can adopt these new and expensive technologies is open to question, although it is being suggested that elements of modern technology, such as electronic chips, can be 'grafted' onto traditional technology.

Finally, the small and medium industries, which are not necessarily in the informal sector, are enjoying a resurgence in developed countries because they and not the large multinational corporations have created most of the needed jobs, and because they are frequently locally owned. Most of the technical innovation is also coming from small firms rather than from multinationals. Furthermore, as technological advances mean that small-scale production is no longer inefficient, small-scale production may be better because of the advantages of ease of management and the ability to change quickly. There is definitely an increased interest in the world in small- and medium-scale enterprises.

Case study F

Small-scale manufacturing in the Philippines

Small-scale industries, defined as enterprises employing between ten and ninety-nine workers, form an important component of the manufacturing sector in the Philippines, comprising 78 per cent of all manufacturing establishments in the country while employing only 18 per cent of the manufacturing workforce (Figure 4.6).

Figure 4.6 The Philippines: role of small-scale manufacturing, 1986

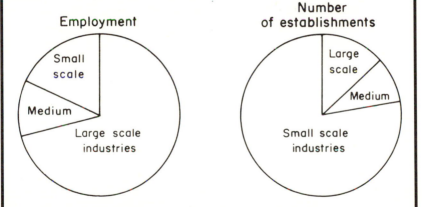

Source: Philippines National Statistics Office (1988) 'Small-scale industries in the Philippines', *Journal of Philippine Statistics* 39:2.

What are the main characteristics of these small-scale industries in the Philippines? Like similar industries elsewhere, they involve owner-operators, who keep in close human contact with their workers and outside businesses. The majority of the owners of small-scale enterprises in the Philippines have either a high school or college education and have frequently completed trade apprenticeships. These entrepreneurs usually start business late in life

Case study F *(continued)*

(after 40 years), indicating the tendency to start a business as a second career. To start and maintain their operations, these entrepreneurs do not raise money on the formal sector financial market, largely because they cannot provide collateral. Many owners still value good social and personal relationships more than the desire to make profits. This tendency to allow one's social obligations and relationships to interfere with the running of the business is frequently cited as a reason for the failure of small-scale enterprises.

Table 4.5 shows the kinds of manufacturing undertaken by small-scale enterprises. Although a wide range of products are manufactured, the majority are engaged in three industries, namely food, textiles/garments, and wood products (46 per cent of the total small-industry employment). The tendency to concentrate on these industries, and the relative absence of heavy manufacturing, is to be explained by reference to the concepts of

Table 4.5 The Philippines: structure of employment in small-scale industries

Type of manufacturing	Number employed	Percentage
Food, beverages and tobacco	25,911	23
Textiles and garments	14,571	13
Leather goods	3,375	3
Wood products	11,269	10
Paper products	2,669	2
Printing and publishing	8,296	7
Chemical products	7,615	7
Rubber and plastic products	7,227	6
Pottery and glass products	947	1
Other non-metallic mineral products	3,434	3
Fabricated metal products	6,617	6
Non-electrical machinery	7,603	7
Electrical appliances etc	3,505	3
Transport equipment	3,383	3
Others	8,274	7
Total	114,696	100

Source: Philippines National Statistics Office (1988) 'The small-scale industries in the Philippines', *Journal of Philippine Statistics* 39:2.

Case study F *(continued)*

Figure 4.7 The Philippines: location of manufacturing value added, 1986

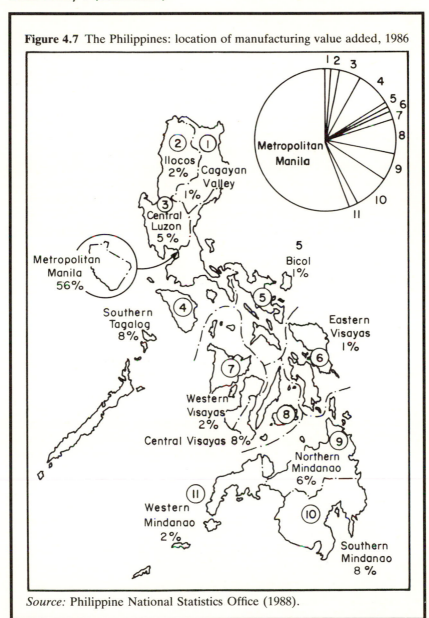

Source: Philippine National Statistics Office (1988).

Case study F *(continued)*

economies of scale and capital intensity. Capital intensity refers to the amount of money needed to start an activity, particularly in terms of setting up the required machinery. In the areas identified above, entrepreneurs can start operations without too much machinery.

The location of small-scale industries in the major administrative regions of the Philippines is given in Figure 4.7, the most striking feature being the overwhelming dominance of the metropolitan Manila area (MMA), which contained over one-half of all manufacturing small enterprises in 1986. The current pattern of location of small enterprises is interesting in that it is very different from that prevailing at the beginning of the decade. In 1980, for instance, MMA accounted for only 3 per cent of all small-scale manufacturing establishments; in 1986, it accounted for an incredible 54 per cent. The biggest loser has been Central Luzon, whose share of establishments fell from 51 per cent in 1980 to 8 per cent in 1986. In general, all areas outside metropolitan Manila have lost small-scale industries. Clearly, a process of centralization has been active in the location of manufacturing enterprises in the Philippines.

Small-scale manufacturers in the Philippines face many problems. Raising starting and working capital is a major problem, as is cumbersome bureaucracy. They also face the problem of poor quality of raw materials and unstable supply schedules. Many entrepreneurs complain of severe competition and of problems with their workforce, such as absenteeism. Nevertheless the Philippines government has been formally assisting small-scale industries since 1966, when it established the University of Philippines Institute of Small Scale Industries with assistance from India. In recent years, the government has given priority to six product groups: garments, furniture, electronics, gifts and homeware, footwear and leather goods, and fresh and processed foods. Small-scale industries had been successful producers of these goods in the past.

The government assists small-scale industries in many ways. Acknowledging their problems in raising money, the government

Case study F *(continued)*

---

both lends through the Development Bank of the Philippines and guarantees loans through the Industrial Guarantee and Loan Fund. Technical services are also provided, such as assistance with feasibility studies through the Small Business Advisory Centres, or by over-the-counter assistance and through a network of extension officers from the Department of Trade and Industry.

(This case study is based largely on Philippines National Statistics Office (1988) "The small-scale industries in the Philippines", *Journal of Philippine Statistics,* 39:2.)

Key ideas

1 Multinational corporations are active in most developing countries, and act as a source of capital, technology, managerial expertise and markets. However, they also present problems since their objectives may not complement national development goals.
2 The state is an important owner of manufacturing in the Third World, although there is increasing pressure on it to divest its involvement.
3 The environment in which manufacturing enterprises operate in the Third World is very different from that of developed countries. Many firms are private family holdings rather than the more familiar western corporations. Quite often there is a high degree of concentration in manufacturing because of the small domestic market.
4 The informal sector still plays an important role in manufacturing in developing countries in terms of both employment and output. It is especially important in utilizing local materials and entrepreneurship, and in promoting balanced regional development. Small enterprises have been especially successful in traditional craft areas.
5 Governments have provided many types of assistance to small-scale industries, including financial assistance, extension support and marketing assistance.

5
The location of manufacturing in the Third World

Introduction

The location of manufacturing activities is an important component of the manufacturing system. Geographers, as we have already noted, have long been interested in analysing the patterns of industrial location. The location of manufacturing activities influences their competitiveness and future viability; it is, therefore, of considerable significance to entrepreneurs that the best location be chosen. In recent years, governments throughout the world have become more aware of the spatial dimensions of development and many have instituted programmes of regional development to reduce spatial inequalities. The location of economic activities, including manufacturing, and the linkages between them, have therefore become of concern to governments too.

The location of a manufacturing establishment is a complex process and involves an evaluation of the relative advantages and disadvantages of potential locations, both at the time of the proposed location and in the future. Because of our inability to gauge the future accurately, locational decisions are clouded by uncertainty. The ability to decide between competing locations depends on the amount and nature of information available to firms and their ability to deal with the information adequately. Since firms vary in this capacity, and given their uneven size and resources, there is scope for suboptimal locational decision making by firms, which subsequently leads to locational movements.

Industrial location can be seen at different levels. First, at the global level, for multinational corporations wishing to invest directly abroad, there is the question of which country to choose among the many potential destinations. Second, once this decision has been made, it has to be decided in which region of that country investment ought to be located. Finally, the MNC faces the micro-level decision of where within the region or the city the facility should be located.

The importance of foreign investment has been steadily increasing in the 1980s as developing countries, weighed down by unprecedented levels of foreign debt, try to earn more foreign exchange and generally expand their economies. This chapter will begin with an analysis of factors that broadly determine the choice of countries for foreign investment. It will then look at the issue of location within countries, identifying the main locational considerations. Finally, it will describe the overall pattern of industrial location in the Third World.

International restructuring and foreign direct investment in developing countries

It has become common now to speak of global economic restructuring. The British geographer, Peter Dicken, has explained it well in his book *Global Shift: Industrial Change in a Turbulent World*. Essentially, during the last three decades, the old pattern of international specialization of economic activities, called the International Division of Labour, has been changing because Third World countries, for long suppliers of raw materials to developed countries and buyers of manufactured goods from them, have been industrializing. They have been able to export manufactured goods to developed countries, making some industries in the latter uneconomic, and forcing their closure. There were other factors that were creating hardships for some industries in developed countries. Labour costs in developed countries were high, and corporations were unhappy with labour militancy. Many corporations, induced by tax and other incentives offered by developing countries, decided to relocate their production there, again leading to factory closures in developed countries. These two factors combined to produce economic restructuring both in developed countries and at the global level. Local restructuring involved the closure of those industries where developed countries could not successfully compete with the newly industrializing countries, and a concentration on new industries in which they had greater competitiveness. In some instances, after a period of suffering because of cheap imports, some industries in developed countries also became more competitive by shedding excess labour, adopting better management techniques, modernizing machinery, and increasing the degree of mechanization.

At the global level, restructuring involved the gradual relocation of many types of manufacturing from developed to developing countries. The types of industries which shifted were initially only the labour-intensive ones, such as garments and cheap electronics. However, as some developing countries established their industrial credibility, with experience and highly educated, relatively cheap labour forces, more complex production was also transferred to them. A good example of what has happened is shown by the activities of the British computer manufacturer, AMSTRAD (see Case study G).

These changes are, then, what are implied by the term 'global economic restructuring'.

Case study G

International economic restructuring and AMSTRAD computers

AMSTRAD was established in only 1968, and initially manufactured home computers. Today, it is the 276th largest industrial company in Britain and has captured a large share of the British and European personal computer market. A major part of AMSTRAD's success has come from producing good-quality, but cheap machines. Being a manufacturer based in a high labour cost country, this posed problems. AMSTRAD met this challenge of producing high-quality low-cost computers by keeping the central component of its operations in Britain, including the design of the computers and chips, and locating its manufacturing operations in developing countries. The company thus takes advantage of skilled British manpower and a long-established scientific and engineering tradition. Keeping the central operations in Britain also means that AMSTRAD is able to quickly assess changes in consumer tastes and requirements as well as keeping abreast of developments in rival corporations.

For manufacturing, which can be undertaken with moderately skilled manpower, AMSTRAD chose to concentrate on countries

Case study G *(continued)*

> such as South Korea (which itself has many local firms producing computers), Singapore and Hong Kong. All these countries are capable of providing not only cheap labour, but also high-quality work. In other words, AMSTRAD did not choose other developing countries which might have offered much lower wage rates, but which would not have provided high quality in production. AMSTRAD meets its tough quality standards by closely supervising and monitoring production.

Factors affecting investment in the Third World by multinational corporations

There are many factors that influence the locational decision making of transnational corporations (TNCs), and the decisions vary with different types of production. In addition, TNCs are extremely large organizations; quite often they will undertake investments that are not immediately profitable and therefore not immediately 'rational', but which are based on other corporate goals, such as creation of market share or reduction of competition. The main factors that determine the location of foreign investment are discussed below.

Political stability   The assurance that they will be allowed to operate their business over a reasonable period of time without turmoil, and that profits and initial capital could be repatriated without difficulty, are of extreme significance to foreign investors. This does not mean that foreign investors will not invest at all in countries in turmoil; where investments are small, profits can be realized in a short period of time, or where extraordinary profits seem possible, foreign investors will still invest. However, for the long-term development of the Third World, these types of investors are not useful.

Favourable and stable public policies   Foreign investment flows to those countries that have policies favouring it, or at least not working against it. Whatever the policies, firms prefer stability. Governments that adopt sound policies and retain them for reasonable periods of time, and those that alter them through dialogue with the business community, are favoured by foreign investors. *Ad hoc* changes and changes for short-term political advantage, so common in the Third World, deter foreign investors.

Large domestic market   In the past, when most developing countries had import-substituting industrialization, foreign investors were attracted by countries that had large domestic markets. As countries have switched to export-oriented industrialization, or in any case have tried to steer foreign direct investment into export manufacturing, the size of the domestic market is no longer that important. However, the domestic market still remains important for many investors who see the prospect of being allowed to produce for it even if, initially, they are permitted to produce only for the export market. The very great interest shown by foreign investors in China in recent years is in part related to its seemingly inexhaustible domestic market.

High levels of infrastructure   On the whole, foreign investment flows to countries where the physical infrastructure is well developed. Physical infrastructure, such as roads, ports, the telephone system, sewerage and power systems, is taken for granted as a precondition for investment. Communication systems, including telex and facsimile facilities, are regarded as critical, as are well-developed air transport networks for the speedy movement of people and goods, especially in the case of high-technology products.

Cheap labour   Cheap, educated and non-militant labour is an important attraction for foreign investors. On the whole, labour costs in developing countries are much lower than in developed countries (Table 5.1). In 1980, for instance, wage costs in Hong Kong, a relatively high wage cost country by Third World standards, were less than 40 per cent of US costs, while Indian wage costs were only 7 per cent of US costs. Within the Third World there is considerable variation in labour costs, which forms the basis of competition among them for foreign investment. However, it is not just the cheapness of labour that is important; corporations look for labour that is educated, experienced and docile too.

Incentives   Most governments offer incentives to both foreign and local manufacturers. The types of incentives offered will be discussed in more detail in Chapter 6; here we can note that there are many types of incentives provided to attract foreign investors, but many researchers

Table 5.1 Labour costs in selected developing countries compared to USA, 1980

Country	Wage cost as a proportion of US wage costs (%)
Hong Kong	38
South Korea	33
Taiwan	26
Singapore	21
Philippines	13
Indonesia	12
Thailand	8
India	7

Source: Adapted from J. Henderson and A. J. Scott (1987) The growth and internationalization of the American semiconductor industry: labour processes and the changing spatial organization of production, in M. J. Breheny and R. W. McQuaid (eds) *The Development of High Technology Industries*, London: Croom Helm.

have argued that incentives do not play that important a role in decision making by TNCs. What is reasonable to assume, however, is that where two competing countries are roughly equal in other attributes, incentives may make the difference as to where the investment goes.

Access to foreign markets An important locational consideration in investment in developing countries by TNCs is whether these countries offer any particular access to markets in developed countries. Many developing countries have preferential market access to developed country markets. For instance, African, Caribbean and Pacific countries have preferential market access to the European Community markets under an agreement known as the Lomé Convention. Thus foreign investors will locate in these countries for this market access. Similarly, the small countries of the South Pacific have duty-free access to Australian and New Zealand markets, whereas Asian and other goods face duty and quota restrictions. As a response to this facility, many Australian and New Zealand manufacturers have relocated their production to the Pacific Islands, which provide cheap labour, tax-free holidays for up to 13 years, and duty-free access to the Australian and New Zealand markets. The importance of market access as a locational consideration has not been fully realized by many writers on location.

So far, we have concentrated on the locational considerations determining to which country investment goes among competing countries. The next section concentrates on factors that determine where in a country manufacturing location occurs.

The main domestic locational factors

Factors affecting industrial location in the Third World

The extent of the locational search (the process of weighing alternative locations) depends on many factors and is influenced by the scale of investment. On the whole, bigger investors conduct more thorough locational searches. Many small, local entrepreneurs do not conduct such lengthy locational searches; they locate in the region of their origin. These small entrepreneurs are also averse to risks and therefore prefer the most familiar and least risky locations.

The main factors that determine the location of manufacturing are as follows (Figure 5.1):

1 Transportation influences the location of manufacturing in terms of transfer costs of raw materials and finished goods, the appropriateness

Figure 5.1 Factors affecting the location of manufacturing in the Third World

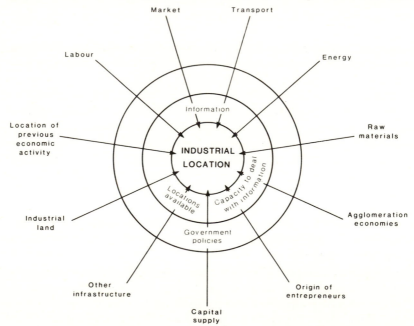

of the mode, and speed. If the manufacturing activity depends on bulky raw materials that are also of low unit value, and if the manufacturing process involves a major reduction in the bulk of raw materials, then the source of raw materials, if produced within the country, or the port of entry, will greatly influence the location.

The second element of transportation cost is the transportation of finished goods. If the market is domestic, as has been the case with most Third World manufacturing, then this will be an important consideration in locational decision making. As we will see later in this chapter, this has resulted in the location of a large share of manufacturing in developing countries in only a few large urban centres. If manufacturing is export oriented, then the port or the airport may be important. In general, the transfer costs of raw materials are much lower than those of the finished product, and this usually means that manufacturers prefer to locate close to the market.

2 Labour is one of the three key factors of production (the others being land and capital) and exercises an important consideration in location. Labour's influence on location will depend partly on the level of skill of the labour force; where high levels of skill are needed, and where these skills are scarce, location will require large labour concentrations and/or provide special inducements for labour. Very few firms actually train their own labour force entirely. Most frequently, the firms rely on an existing labour pool. This means that establishments will tend to locate in countries that already have an industrial labour force and in or near urban centres.

But the supply of labour in terms of its physical availability is only one of the labour-related considerations. There are frequently variations in the quality of labour in different parts of the country, and location will normally take this into account. Moreover, the rates of compensation will differ to some extent by location, and manufacturing activities in which labour forms an appreciable portion of the total cost of production will often require cheaper labour localities. Thus new manufacturing establishments will often search out suburban locations, or tend to locate in the smaller cities and towns. This often represents an important inducement for economic decentralization.

3 One of the critical locational factors is obviously industrial land. Manufacturers need well-serviced land in sufficient quantity at reasonable prices. This need is often not met in the Third World, and where it becomes available this can be an important locational pull. Land for industrial purposes is usually zoned away from residential

uses and is available as land for other purposes. However, increasingly, in an attempt to rationalize manufacturing location and to induce and encourage greater industrialization, industrial estates have been provided. These estates are simply special areas marked for industrial use that the government, statutory agencies and private corporations provide, and often include completely serviced sites, in many cases with buildings already erected.

4 All economic activities need basic services before they can function properly. Infrastructure refers to those facilities and services that are provided communally for the use of everyone, and are frequently too expensive to be provided by any one manufacturer. In addition to the requirements of infrastructure, firms will need producer services, such as banking, wholesale services, communications, legal services and so on. Often the largest urban centres have this basic infrastructure, and this frequently attracts new manufacturing there.

5 Energy is a variable but important component of industrial location. In some industries, such as metal-based industries, energy is a very important consideration in location. Some regions or countries sometimes sell energy at concessional rates to attract certain types of manufacturing.

6 Manufacturing often develops in the Third World from other activities, quite frequently from wholesaling and retailing. The new manufacturing activities are then simply added to the main location. There is a desire to keep the business together, and since the family runs the business to keep it together too. It has been observed that in most developing countries, entrepreneurs locate their manufacturing where they reside at the time an enterprise is formed. This tendency to locate in areas of birth or residence has been observed in many developing countries, and in fact has been shown to be true of Britain too – in the case of family-owned manufacturing.

7 Capital is one of the critical elements of production and exerts an important influence on location. Some areas will have much better institutions for the supply of credit – such as large banks, trust companies and insurance corporations – that will pull in new manufacturing firms.

8 Nearly every economic activity depends on benefits which accrue without expenditure. These are benefits that accrue to firms because of the agglomeration of many other firms in the same location. Benefits such as the presence of infrastructure, the ability to draw from a common labour pool, the ability to minimize the cost of

information flow, and generally the ability to draw on required services, are all the result of the increased size of the centre in which the particular establishment is located.

9 The influence of the government, which will be covered more fully in Chapter 6, extends to the location of manufacturing establishments which it influences in many different ways. First, as we have seen, the government can and does influence the location of manufacturing activities through the use of industrial estates. In recent years, governments have developed export processing zones (EPZs) to encourage export-oriented manufacturing. Where these EPZs are located greatly influences the locational pattern of manufacturing. More generally, the government's pattern of investment in infrastructure will also influence location. Sometimes, where the government's industrial policies show sophistication, the government will use selective inducements for locating in certain areas. Thus it may designate backward areas, or depressed areas, and give special inducements if firms locate there.

The pattern of location of manufacturing in developing countries

The most outstanding feature of the location of manufacturing in developing countries is its very high degree of concentration in the largest urban centres, or in just a few places. In Chapter 3 we saw the high degree of spatial inequality in the distribution of manufacturing between countries; the same tendency towards spatial concentration occurs within developing countries. Whilst information is sparse on the location of manufacturing in the Third World as a whole, sufficient data exist on some countries to illustrate this tendency towards spatial concentration (see Case study H).

Case study H

Location of industry in South Korea

In a recent survey of the manufacturing system of South Korea, Park reported that 96 per cent of all headquarters separated from their production units were in the Seoul metropolitan area, a rate much higher than in Japan and other developed countries. Although the distribution of factories (as opposed to headquarters

Case study H *(continued)*

of firms) shows slightly less concentration, Seoul clearly dominates the manufacturing sector. Table 5.2 shows the distribution of branch offices of manufacturing firms in Park's study.

Table 5.2 South Korea: location of branch offices of manufacturing firms

Location	Percentage of branch offices
Seoul	35
Pusan	20
Taegu	11
Kwanju	7
Taejon	7
Wonju	6
Chonju	5
Masan	3
Kyongju	2
Other cities	4

Source: Sam Ock Park (1986) 'Regional changes in the industrial system of a newly industrializing country: the case of Korea', in F. E. Ian Hamilton (ed.) *Industrialization in Developing and Peripheral Regions*, London: Croom Helm.

Notwithstanding this concentration, one of the major changes in South Korean manufacturing in the last two decades has been its diffusion from Seoul to other areas, particularly to Pusan, where much of the heavy and chemical industries have been established. In the late 1970s and the 1980s, a large proportion of new industries being set up were established in the smaller cities and rural areas within the Seoul metropolitan area. During 1979–81, for instance, Seoul accounted for 35 per cent of all plant closures and 9 per cent of plant additions of the thirty-two groups studied by Park.

Some of the locational changes have been due to public spatial policies, particularly the Local Industrial Development Law introduced in 1971 and the First Ten Year Comprehensive National Land Development Plan, covering 1972 to 1981. However, South Korean economic restructuring, itself a response to changing international processes, has also been responsible for locational change.

In some recently reported research, it was found that 51 per cent of the high-technology firms in Israel were located in Tel Aviv metropolitan area, and that the Tel Aviv and Haifa corridor accounted for the location of 62 per cent of these high-technology firms. In the case of India, a country which has engaged in detailed economic and industrial planning, two states account for nearly 40 per cent of manufacturing employment, and six states account for over 70 per cent of all manufacturing employment: Maharastra, West Bengal, Gujarat, Tamil Nadu, Bihar, and Utra Pradesh. In peninsular Malaysia, of the three main regions, western states contained nearly 75 per cent of all manufacturing establishments in 1979, while southern states contained 20 per cent and northern states only 5 per cent. In Colombia, industrial development is also spatially concentrated: in 1980, only one region (Cundinamarca) accounted for one-third of Colombia's manufacturing employment, while three regions contained 74 per cent of all manufacturing employment. In Kenya, about one-half of manufacturing establishments producing 55 per cent of the national value added were located in Nairobi; as Table 5.3 shows, this is typical of many African countries.

Table 5.3 West African capital cities' share of manufacturing establishments, 1971

Country	Capital city	Percentage share
Gambia	Banjul	100
Liberia	Monrovia	100
Senegal	Dakar	82
Sierra Leone	Freetown	75
Ivory Coast	Abidjan	63
Guinea	Conakry	50
Nigeria	Lagos	35
Ghana	Accra	30

Source: J. O. C. Onyemelukwe (1984) *Industrialization in West Africa*, London: Croom Helm.

The pattern of industrial location in developing countries, then, is characterized by urban primacy, which refers to a situation where the top-ranking urban centre is overwhelmingly important to the country's economic system. Primacy comes through the process of agglomeration, where a centre with the initial location of economic activities becomes more attractive with further location, which then exerts even more influence on further location; there is a snowballing effect. Gunnar Myrdal, the famous Swedish sociologist and economist, has coined the term 'the process of cumulative causation' to describe this concentration of economic activities.

Apart from the purely economic factors that have led to concentration in industrial location, other processes have been important. First, the fact that import-substituting industrialization was adopted meant that industries tended to be located where the market was, that is in the main urban centres. Second, many governments developed the best infrastructure in the main cities and towns. For instance, the first and best industrial estates were developed in the main centres; these reinforced the concentration of manufacturing in just a few centres. Finally, many governments either did not institute policies to decentralize economic activities or, where they tried to do so, their attempts were feeble.

An example of industrial location is given in Case study I.

Case study I

Location of manufacturing in Fiji

Data gathered during a study in 1983 (Figure 5.2) show the distribution of gross manufacturing output of establishments, while Table 5.4 shows the distribution of the urban population in Fiji, a widely used surrogate for the distribution of manufacturing activities. No matter how one examines the spatial distribution of manufacturing in Fiji, it is clear that Suva dominates the manufacturing sector, containing about one-half of manufacturing employment. Suva, of course, is dominant not only as a manufacturing sector; it is a primate city, dwarfing all other urban centres in the location of population, economic activity, political power and general infrastructure. Lautoka is the second most important industrial city, having developed from its sugar milling origins, and now includes some of the most dynamic manufacturers. Ba and Labasa, the regional capital of the other main island, are the only other major industrial centres; the rest of the urban centres are insignificant in industrial production.

So what explains Suva's overwhelming domination of manu-

Case study I *(continued)*

Figure 5.2 Fiji: location of manufacturing gross output, 1983 (F$000)

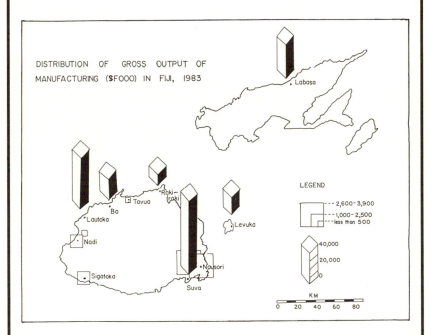

DISTRIBUTION OF GROSS OUTPUT OF
MANUFACTURING ($F000) IN FIJI, 1983

LEGEND

---- 2,600–3,900
---- 1,000–2,500
--- less than 500

40,000
20,000
0

KM
0   20   40   60   80

facturing in Fiji? One of the main reasons is to be found in its colonial heritage. Colonial development emphasized the growth of only a few areas; the government was not interested in economic decentralization. Furthermore, Fiji's exports of commodities did not generate enough surplus for the government to undertake equalization programmes, even if it wanted to do so. Suva also benefited from being the centre of European settlement; since Europeans controlled both private and public investment, it may have received particular attention. The market economy similarly magnified Suva's attractiveness once the first economic activities had been established there. Most of the industrial land was also developed in Suva, which was seen by the government as its

Case study I *(continued)*

showcase to the outside world. Finally, foreign investment was attracted to Suva not only because it possessed the largest port and the best infrastructure, but also because it afforded the best access to political power.

Table 5.4 Fiji: urban hierarchy, 1986

Urban centre	Population	Rank	Expected population	Deviation (% of ideal population)
Suva	141,273	1	–	–
Lautoka	39,057	2	70,637	55.3
Lami	16,707	3	47,091	35.5
Labasa	16,537	4	35,318	46.8
Nadi	15,220	5	28,255	53.9
Nausori	13,982	6	23,546	59.4
Ba	10,260	7	20,182	50.8
Vatukoula	4,789	8	17,659	27.1
Sigatoka	4,730	9	15,697	30.1
Rakiraki	3,361	10	14,127	23.8
Levuka	2,895	11	12,843	22.5
Savusavu	2,872	12	11,773	24.4
Navua	2,775	13	10,867	25.5
Tavua	2,227	14	10,091	22.1
Korovou	340	15	9,418	3.6

Source: Calculated from data in Fiji Bureau of Statistics (1988) *Report on Fiji Population Census 1986*, Vol. 1, Suva: Government Printer.

Micro-level locational factors

So far we have examined industrial location in Fiji at the national level; but what about micro-level locational processes? A multiplicity of factors determine industrial location at the micro-level. A study in Fiji in 1983 obtained detailed information from manufacturers on factors determining their micro-locations (Table 5.5). Of these the single most important reason given for location was the prior ownership of properties, particularly among locally owned firms. Thus in the case of locally owned manufacturing establishments, the residence of the owners will largely determine the location of plants. This pattern can be seen very widely in Fiji. Punja and Sons, for instance, has planned most of its

Case study I *(continued)*

manufacturing establishments in Lautoka, the home town of the firm. This tendency to locate manufacturing establishments in the area where the owner was born, or where he/she has other interests, is a well-recognized feature of industrial location in the Third World.

Table 5.5 Fiji: reasons for location of manufacturing establishments, 1983

Reason	Number of establishments	Percentage
Owner(s) already had property	19	25
Only subdivision or industrial zone available at the time	13	17
Customer accessibility	9	12
Large tract of suburban land available	6	8
Reasonable price of property or reasonable rent	6	8
Proximity to port	5	7
Land given by government or location required by government	4	5
Right size of property	4	5
Freehold land	3	4
Not congested	2	3
Good subdivision	2	3
Sea frontage	1	1
Close to power supply	1	1
Raw material location	1	1
Not available or property acquired with purchase of firm	32	–
Total	108	100

Source: Survey of manufacturers in Fiji, 1983.

Not surprisingly, the second most significant micro-locational determinant was the availability of only one industrial subdivision or industrial zone when the firm in question was being established. This finding, too, reinforces our earlier observation that industrial estates have exerted a major influence on industrial location in Fiji.

The importance of customer accessibility is explained by the importance of consumer orientation of manufacturing, and relates particularly to tailoring and furniture making. The remaining

Case study I *(continued)*

reasons are diverse. The impression that emerges from Table 5.5, and from interviews with manufacturers, is that the choice of industrial location has been severely limited in the past and that many firms have not undertaken analyses to determine their best location. Of course, in an environment in which feasibility studies are often absent and frequently inadequate, it may be inappropriate to expect locational analyses.

Locational stability and mobility in Fiji

Manufacturing establishments in Fiji have been fairly stable in their location; only 4 per cent had been at the current location for 1 year or less. There are some establishments, mainly sugar and copra mills, that have been operating for long periods. Nearly one-half of the establishments in the sample had been at their locations for less than 10 years, indicating the recency of industrial expansion in Fiji. Another indicator of locational stability in the

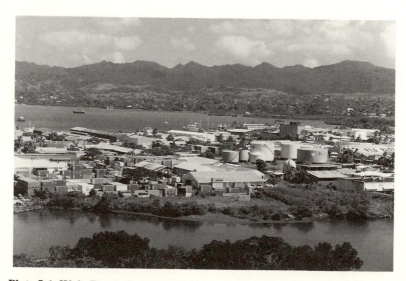

Plate 5.1 Walu Bay Industrial Estate, the first one to be established in Fiji.

Case study I *(continued)*

manufacturing sector was in fact that the majority of manufacturing establishments (60 per cent) had not made any move from
where they were first established. This locational stability does not
mean that stability will continue. Indeed, 30 per cent of the
establishments indicated that moves were planned within the next
5 years, while 65 per cent did not plan to change location. Six
per cent of the establishments were unsure of their locational
intention.

Plate 5.2 Women workers in a noodles factory in Suva, Fiji. Women
comprise a large proportion of the manufacturing labour force in Fiji.

The data indicate that a major locational adjustment is indeed
in the offing. Most of the proposed changes result from a period
of *ad hoc* growth on the part of firms, which are now moving to
rationalize their operations. Thus 40 per cent of the establishments
planning to relocate within the next 5 years said their move was
motivated by expansion of their operations. For the remaining 60
per cent, the reasons for planning a move varied widely. These

Case study I *(continued)*

included the purchase of properties by the firms and the desire to move into them (in the case of small firms); moving to locations with greater consumer accessibility (mainly furniture makers and tailors); being forced out of residential areas; the desire to move from native lease locations to others, preferably freehold; and the avoidance of floods.

Nearly one-half (45 per cent) of the establishments contemplating relocation did not know where they would move to. For those that knew where they planned to move, nearly always the new location was a suburban one, frequently in emerging industrial estates. Namaka in Nadi, and Navutu in Lautoka, were popular destinations, mostly because these new areas offer the only available industrial lots in their respective areas, and because they offer prospects of large lots. Within Suva itself, planned moves were to Walu Bay, Kalabo and Vatuwaqa. Again, the last two represent new industrial estates – and thus the only possible choices

It becomes quite clear that most establishments are planning to remain in the same area; they are merely changing their sites as it were. In only one case was the planned movement to a distant location. Fiji is unlikely, therefore, to experience any major shift in its present pattern of industrial location in the foreseeable future.

Key ideas

1 The location of manufacturing activities is part of a firm's overall strategy and involves the evaluation of many factors without knowing the likely future changes, and without possessing all relevant information.
2 Multinational corporations conduct extensive locational analyses before deciding in which country to invest. Although many factors affect the decision of transnational corporations, domestic market size and foreign market access, security of capital, and favourable and stable government policies are most important.
3 Within a country, a host of factors affect a firm's locational decision,

although the overriding consideration is to choose a location that yields the greatest revenue. However, many small entrepreneurs locate their manufacturing activities where they reside at the time of formation of the firm, or close to their other economic activities. Thus many of the factors that are assumed by various theories to determine location do not operate.

4 Governments influence industrial location both by intervening in the cost of inputs (labour costs, electricity, land, transport) and by specific locational policies (such as the provision of industrial estates and export processing zones and location-specific incentives).

5 Manufacturing in developing countries shows a high degree of spatial concentration in the main urban centres, this being a reflection of the wider process of economic inequality.

6
The state and industrialization in the Third World

Introduction

The state plays a crucial role in practically all economies, providing an administrative and legal framework for economic activities, social and economic infrastructure, and incentives and disincentives (through monetary, fiscal and exchange rate policies) to guide the allocation of resources. In addition, in most economies the state is also assigned the role of protecting the environment and ameliorating economic inequalities. It also often plays a direct role in production but does not always pursue policies conducive to rapid industrialization and development, as we will see from the case study of Togo. States can undermine economic efficiency through excessive bureaucratic controls, nepotism, ethnic favouritism and corruption. However, the role of the state is of crucial importance in Third World development, and it has been receiving increasing attention in the recent past because of the magnitude of direct government production and the increasingly large budget deficits experienced by governments.

The state has always been closely involved in industrialization. In the case of presently industrialized market economies, it provided protection to nascent industry. As we saw in Chapter 2, most of these governments also engaged in colonialism, providing their industries with investment and much needed raw materials and captive markets. The current international financial and trading system was also put in place by these countries; these arrangements have served the interests of their industrialists very well. Hence the call by poor countries for a New International Economic Order.

The role of the state in global economic activity has increased dramatically in the last century, first in developed countries from 1880 onwards, and in developing countries from 1940 onwards (World Bank 1988: 5): in 1880, the average (unweighted) share of government expenditure in the GDP of six developed countries (France, Germany, Japan, Sweden, United Kingdom and the United States) was about 10 per cent; in 1985, it had reached 47 per cent. The share of government expenditure in developing countries is much lower, being on average just above one-quarter of GDP. However, in nearly all cases, the financial difficulties of the 1980s have led to a reduction in government expenditure. In the developed countries governments have been reducing their limited social welfare programmes, as well as selling assets, while in developing countries, assets have been sold through a programme of privatization, a topic dealt with in Case study J (see pp. 103–4).

This chapter will first look at the different roles played by the state in industrialization in developing countries. It will then examine the two main types of industrial policies that governments have used, namely import-substitution and export-oriented industrialization. A case study of inappropriate industrial policies and state enterprises in Togo, a small West African country, will illustrate some of the issues discussed in this chapter.

Different roles of the state

We have already noted the adoption by most developing countries of planning as their route to rapid economic transformation. The state in developing countries plays three main roles as shown in Figure 6.1: facilitative, regulative, and direct production. It provides an overall environment conducive to industrialization by assuring security of investment, providing infrastructure, tax concessions, protection from imports and local competition, subsidies, and by negotiating with foreign governments to improve market access for domestic products. In addition to facilitating industrial production, the state also regulates the sector through land use and other rules that are designed to serve the best interests of society. It typically has regulations covering most aspects of production, including land use, safety and health of workers, use and borrowing of funds, and takeovers. Finally, the state also contributes to the manufacturing sector through direct production

Figure 6.1 The state and industrialization in the Third World

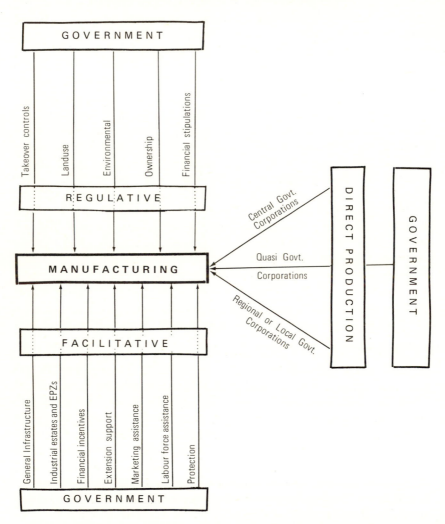

THE STATE AND THE MANUFACTURING SECTOR

by government departments, state-owned enterprises (SOEs), joint ventures, or commercial enterprises in which it holds minority shares. The state can work directly through its central agencies or through its various levels of regional and local governments.

Import-substitution industrialization

The drive for industrialization in developing countries began with import-substitution industrialization (ISI). Figure 6.2 outlines the evolution of industrial policies in the Third World. Almost every developing country, including those now regarded as models of successful industrializers, such as Brazil, Singapore and South Korea, used this strategy. ISI involved the local production of previously imported manufactured goods. Industrialization in developing countries was given impetus by the breakdown in international trade during the Second World War, which necessitated some industrialization in the colonies, by the depression of the 1930s in developed countries, and by the fluctuating or deteriorating prices of agricultural commodities.

Why did developing countries adopt an import-substituting strategy rather than producing for export? There were many reasons. First, post-colonial societies were acutely conscious of depending too much on their previous masters; they wished to become self-sufficient. The ISI strategy

Figure 6.2 The evolution of Third World industrial policies

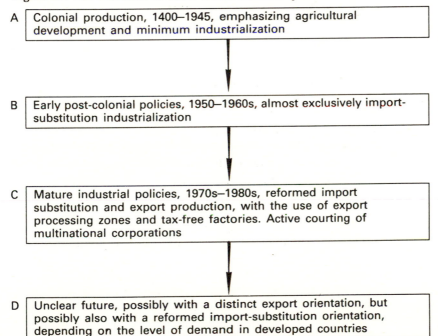

A — Colonial production, 1400–1945, emphasizing agricultural development and minimum industrialization

B — Early post-colonial policies, 1950–1960s, almost exclusively import-substitution industrialization

C — Mature industrial policies, 1970s–1980s, reformed import substitution and export production, with the use of export processing zones and tax-free factories. Active courting of multinational corporations

D — Unclear future, possibly with a distinct export orientation, but possibly also with a reformed import-substitution orientation, depending on the level of demand in developed countries appeared to offer them the chance to reduce their dependence on developed countries by relying on domestic markets and building an indigenous technological capacity. Second, export production was not feasible then because of the inability of developing countries to compete with established producers. Third, ISI allowed developing countries to foster the growth of industrial activities by national entrepreneurs; an export-oriented strategy would have necessarily involved extensive foreign ownership in manufacturing. In order to nurture industries, developing countries provided high levels of protection to domestic manufacturing; these included tariffs on overseas goods and licensing, with imports being allowed only to supplement domestic production. Protection accorded to industry was further enhanced through tough screening of foreign investment, thus limiting competition.

Protection of domestic producers has been only one element of the multi-pronged development and industrial strategy followed by the state in developing countries. Other measures to encourage manufacturing have included a wide range of incentives offered to both foreign and local capital. Perhaps one of the most common attempts to attract capital has been the offer of financial packages. Incentives offered include substantial tax holidays, usually of 5 to 10 years; generous depreciation allowances; loss carryover; reinvestment allowances; and export incentives providing for tax exemption on various proportions of profits derived from exports. Governments have also assured entre-preneurs, both local and foreign, of freedom to move capital, and have also provided the infrastructure. In particular, most developing countries have provided industrial estates to try to eliminate bottlenecks caused by shortage of industrial land.

Extension services have been provided for manufacturers, particu-larly for small-scale units and the informal industrial sector, covering management assistance, machinery checking and maintenance, and quality control. Governments have also helped the manufacturing sector through training, providing broad education, vocational training and on-the-job training. Finally, governments have further assisted the manufacturing sector by providing cheap capital. Sometimes they have provided capital through equity in new ventures, and sometimes through subsidized interest rates, usually through national 'develop-ment' banks. A major component of government policy in developing countries involved the participation of the state in direct production. Because of the extent of its participation, direct production is dealt with in detail below.

Direct production

State-owned enterprises (SOEs) have become an important feature of the industrial landscape in developing countries and their importance has been increasing in the last decade. Although separate data on the role of public enterprises in developing countries are not readily available, it is known that of the 500 largest non-US corporations included in *Fortune* magazine, 71 or 14 per cent were manufacturing SOEs, accounting for 19 per cent of the sales, 21 per cent of the assets, and 26 per cent of employment (not all of these corporations are in developing countries, though). Furthermore, SOEs have been improving their position in recent years; from 1975 to 1984, they improved their shares of sales, assets and employment.

In a survey of eleven developing countries, UNIDO has reported that the average public sector contribution to output was 44 per cent but ranged from 8 per cent (Mexico) to 85 per cent (Somalia). The share of the public sector in investment in manufacturing averaged 48 per cent but varied from 9 per cent (Morocco) to a high of 90 per cent (Egypt). In a more recent survey, the World Bank has reported that of nineteen countries, the majority had an SOE share of investment of over 20 per cent, and that in three countries (Zambia, Burma and Venezuela) the share was more than 50 per cent. The contribution of the public sector to employment is of a similarly large magnitude.

The state engages in direct production for a wide variety of reasons, perhaps the most important being the absence of a domestic middle class that can spearhead industrialization. Thus the state finds little alternative to direct production to spur industrialization and retain local control of resources. In addition, it often seeks to retain control of strategic sectors, such as steel industries, or defence-related aircraft industries. Sometimes certain industries are neglected by the private sector because of high risks, or because of large capital requirements, and the state steps in to fill these gaps. Sometimes states involve themselves in direct production to overcome problems of racial or ethnic inequalities, or to help reduce spatial inequalities. The state may wish to promote small or labour-intensive industries and may engage in direct production to ensure this. Finally, states occasionally need to control the prices of goods, particularly those needed in the production of other goods (intermediate goods).

While the state is involved in nearly every kind of manufacturing, its involvement tends to be greatest in sectors requiring large capital outlays, such as steel and fertilizer industries, or where national security is concerned, such as telecommunications and aircraft industries.

As the state's direct involvement in the industrial sector and in the economy as a whole has grown, a concern for the efficiency of public enterprises has emerged. In most developing countries, given the recency of the phenomenon, organization and coordination in the public sector are poor. There have been queries about the performance of SOEs in relation to private sector companies, especially as most SOEs have enjoyed monopolistic markets and easier access to government funding, and have benefited from many government services and goods. State enterprises frequently became a drain on government resources, needing huge subsidies and contributing little to government revenue (see, for instance, Case study K later in this chapter). In many instances, SOEs contributed to an increase in capital intensity because of their easy access to capital, and thus did not make as big a contribution to employment generation as had been hoped. Moreover, SOEs also contributed to the increasing indebtedness of developing countries. For instance, in Brazil, Mexico, the Philippines, Portugal and Zambia, SOEs account for more than half of all external public debts.

Thus most countries have been attempting to review the position of SOEs, trying to make them more efficient. Some countries, such as Argentina, Guyana, Malaysia and Mexico, have established separate ministries of public enterprises, while others, such as India, have established advisory and supervisory entities. These attempts to improve the organization and efficiency of public enterprises are now permeating the bulk of developing countries, especially under the influence of the World Bank and the International Monetary Fund.

Attempts to empirically evaluate the efficiency of public enterprises have been scarce and the evidence is conflicting. On the whole, public enterprises have been found to be less efficient than private sector firms, partly because of the lethargy of government bureaucracy, partly because of government interference, and partly because of the nature of the industries themselves. For instance, steel industries, whether private sector or publicly owned, have performed poorly. On the other hand, some state enterprises, particularly in heavy engineering and machinery, have been very successful and have achieved impressive export performances. Bharat Heavy Industries of India and Hindustan Machine Tools Limited are good examples of these. In the case of Fiji, the state-owned Fiji Sugar Corporation has not only expanded sugar-cane cultivation and sugar production, but also modernized the mills, paid dividends, and developed subsidiary enterprises utilizing its by-products. It has performed much better than its predecessor, the multinational Colonial Sugar Refining Company of Australia.

Improved performance of SOEs depends on reaching a clear position regarding the role of the government in their establishment and operation; consistent application of state policies; the presence of selective competition; freedom from government controls on price and raw materials; freedom from government in employment and labour processes; and stringent financial monitoring and discipline.

Export-oriented industrial policies

Import-substitution industrialization led to rapid increases in industrial production in most developing countries as both local and foreign entrepreneurs took advantage of government financial incentives and market protection. However, ISI led to the creation of high-cost industries because small domestic markets meant full economies of scale could not be realized. Initially, this was not a major problem because basic food and consumer goods had large markets, but it became a major problem as countries tried to proceed to the second round of import substitution involving more specialized goods which needed large markets for efficient production. Costs also increased because of the absence of competition both from local producers and imports. Entrepreneurs did not therefore strive to reduce costs and improve the quality of their products. In addition, ISI was seen to have failed to reduce external dependence, since in many instances raw materials were imported. More importantly, firms in developing countries had bought or licensed technology from developed countries. Finally, the heavy involvement of the state, particularly through SOEs, was proving to be a major drain on resources.

These shortcomings, the need to earn foreign exchange, and pressure from international agencies, particularly the World Bank and the International Monetary Fund, have led to the adoption of export-oriented industrial policies, particularly since the 1960s. Led by Singapore, Taiwan and South Korea, most developing countries began to orientate their policies to produce for the world market rather than for the often small domestic market. Even if countries did not embrace global production fully, they none the less modified their policies and encouraged exports. In the last few years, as countries have become highly indebted and experienced debt-servicing difficulties, they have also become more vulnerable to the pressure of the IMF and the World Bank to pursue export-oriented policies. The change towards export-oriented manufacturing has been achieved with the policy changes outlined in Table 6.1. A key element in the promotion of export-oriented industrialization has been the use of export processing zones, dealt with in more detail in the next section.

Table 6.1 Policy changes used to achieve export orientation in Third World manufacturing

A	Exchange rate adjustments, including devaluations, to make goods cheaper in international terms
B	Active search for markets, including conclusion of trade and aid agreements, organization of overseas trade exhibitions, and market intelligence
C	Open trade policies, with a reduction of import protection and increased competition in domestic manufacturing
D	Provision of foreign exchange risk protection
E	Discriminant tax and duty benefits favouring exports
F	Reduction of tax on exports
G	Provision of export processing zones, which cover a package of financial benefits and streamlined bureaucracy
H	State intervention in the labour market to ensure the supply of docile, non-unionized, cheap labour for export production
I	Relaxation of laws regarding ownership and local borrowing
J	Privatization of state activities to reduce the role of the state in direct production and generally to increase the role of the market in the economy

Export processing zones (EPZs)

The establishment of export processing zones is a fairly recent phenomenon, in a way being an extension of the industrial estate. Export processing zones comprise not only a spatial entity – an (often) enclosed area within which firms are given a wide range of benefits, movement to and from which is closely monitored – but also a package of incentives. In some countries, such as Mauritius, the whole country is deemed to be an EPZ; that is, firms are not required to locate in a special zone to benefit from incentives. However, to be allowed into an EPZ, firms have to export a high proportion of their output.

There has been a dramatic expansion of EPZs in the world since their introduction in Costa Rica and India in mid-1960. In 1986, forty-eight countries in Africa, Asia, Latin America and the Caribbean operated 180 EPZs; and a further eighteen countries offered similar conditions without formal EPZs. Furthermore, twenty-two countries had EPZs under construction. Workers in EPZs and offshore factories comprise about 5 per cent of the manufacturing labour force of developing countries.

Figure 6.3 Third World export processing zones

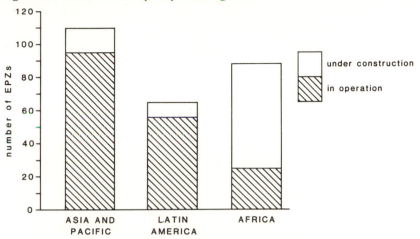

Source: Kreye, O., Heinrichs, J., and Frobel, F. (1987) *Export Processing Zones in Developing Countries*, Working Paper No. 43, Multinational Enterprises Programme, ILO, Geneva.

The distribution of EPZs is shown in Figure 6.3; it clearly reflects the relative position of these regions and countries in export-led industrialization. Africa stands out with the fewest EPZs; it is also the least industrialized region and one that has not experienced much industrial growth in the last two decades. Asia and the Pacific region, particularly South-East Asia, contains the largest numbers of EPZs; this region has also experienced dramatic success in export-led industrialization.

Export processing zones represent the most dramatic manner in which developing countries are competing for foreign investment and trying to induce local manufacturers to produce for exports. A wide range of incentives are offered as part of an EPZ package. Although details of incentives often vary from country to country, the general types of incentives offered are extensive (see Figure 6.4). Export processing zones engage in a wide range of manufacturing, but they have typically concentrated on the textile and garment industries and electronics, largely because cheap labour is important in these industries since mechanization has not been feasible. On the whole, the EPZs have employed young female labourers.

Although EPZs have led to an increase in exports, they have generated controversy as a development strategy. Activities in these

Figure 6.4 A typical EPZ incentive package

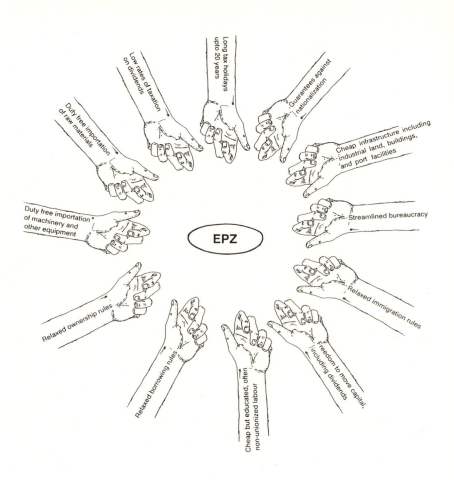

zones have generally not developed strong links with the rest of the economy; the economy becomes more dependent on overseas markets; wages are low; women are 'retired' at a very early age (sometimes by 25 years); and there is hardly any transfer of technology. On the other hand, these zones are seen as attracting additional investment, earning foreign exchange and providing employment.

Case study J

Privatization

A major feature of export-led industrialization has been the growing degree of privatization. The problems of state-owned enterprises, external debt-servicing difficulties and pressure from international financial organizations have led almost to a privatization revolution in the world, including the Third World. Based on the Thatcherite belief in limited government, Britain provided the lead in privatization, transferring a third of its nationalized workforce to the private sector. Other developed countries followed suit, such as France, which had previously nationalized a large number of industries as early as 1981. As well, the Soviet Union and China, for so long flagships of the socialist model of development, began to transform their social and economic systems to include more private production and distribution. In the case of China, however, recent events in Tiananmen Square, when the student-led revolt was crushed brutally by the army, have cast doubts on its long-term commitment to socio-economic reforms.

Privatization can be achieved in many ways, including the sale of public corporations to employees, the public, or to other corporations; the conversion of SOEs into private-sector-type organizations with a commercial mandate; the contracting out of services previously provided by the state; the use of private management in state-controlled enterprises; and the use of user fees in services. In addition, governments have tended to reduce protection for SOEs, forcing them to compete with private local and foreign corporations.

Like developed countries, Third World countries have also embraced privatization (Table 6.2). Those countries that most dramatically privatized their economy are also heavily indebted, such as Brazil, Mexico and Togo (see Case study K). For these countries, privatization is not a carefully thought-out philosophical change, but an urgent practical necessity: they no longer have the money to subsidize their enterprises and cannot obtain new loans until they agree to the reforms proposed by international financial institutions. Some of these countries are also using privatization to reduce their external debt through complex debt-for-equity-

Case study J *(continued)*

swapping arrangements. However, even countries not immediately under financial pressure are privatizing to reduce their budget deficits, and generally to introduce dynamism in their economies.

Privatization can, however, create major problems for developing countries by significantly raising the level of foreign ownership of the economy, and by preventing governments from being able to redistribute incomes to disadvantaged groups and regions. Furthermore, there is no clear-cut evidence yet that less government necessarily means better or faster industrialization. A case can be made that Taiwan and South Korea, which have performed exceedingly well in industrialization, show that a well-tuned state interventionist policy is important in industrialization.

Table 6.2 Privatization in selected developing countries

Country	Details of privatization
Brazil	Sold 12 state corporations in 1985 with a further 77 for sale
Bangladesh	More than 600 state firms have been denationalized since 1975
Chile	Of 377 state enterprises in 1973, only a few remain today; most have been sold to banks and insurance companies
Costa Rica	More than 700 corporations have been sold to employees
Guatemala	About 31 state firms have been sold to employees
India	Private buses have been allowed for the first time in Calcutta
Mexico	Put 236 firms up for sale in 1985 with a further 80 to be sold later, including the national airline
Somalia	In 1982, 110 government-owned boats were sold to private entrepreneurs
Togo	Began privatization in 1983 with the sale of 5 manufacturing corporations (see Case study K)

Source: Randall Fitzgerald (1989) 'The privatization revolution', *Economic Impact* 67:35.

Import substitution or export orientation?

Import-substitution policies have been severely criticized not only by academics of all persuasions – including dependency writers – but by many developing country analysts themselves, and by the international financial community, particularly the International Monetary Fund and the World Bank. Critics have argued that import-substitution policies have produced an inefficient industrial structure that has been incapable of solving the unemployment problems of developing countries. As a result, these policies have been modified, their excesses eliminated, and a strong export orientation has been introduced. The current thinking strongly favours export-oriented industrialization.

On balance, international experience has firmly indicated that excessive inward-looking policies inhibit development in the long run because domestic economies are deprived of a powerful source of information, technology and, most importantly, competition. Furthermore, many developing countries have small domestic markets, and an import-substitution policy imposes severe penalties of economies of scale. The experience and trade performance of those countries that have followed import-substitution industrialization, such as India, have been less impressive than those of countries, characterized as NICs, such as South Korea, Taiwan, Singapore or Hong Kong, that have followed an export-oriented industrialization strategy.

Having said this, however, it must be pointed out that import-substitution industrialization has served an important purpose in developing countries. Without it, industrial development would have been minimal or non-existent. Furthermore, import-substitution industrialization helped prepare an industrial base for successful forays into exporting. Countries such as Singapore, South Korea, Taiwan and Brazil, all countries regarded as having been successful industrializing countries, began their industrialization with import substitution. It has not been shown by critics of import-substitution industrialization that these, or other countries, could have become successful exporters of manufactured goods without the assistance of import-substitution industrialization in the initial stages of their industrialization programmes. The real failure of import-substitution industrialization lay not in the strategy itself, but in the inability of these countries, save for countries such as Singapore, South Korea, Taiwan and Brazil, to review their policies critically to reduce their high levels of protection to domestic manufacturing industries, and to adjust their incentive systems to remove disincentives to exporting.

The developing countries thus need to formulate their industrial strategies with both the domestic and international situation and prospects in mind, and to allow as much exposure to international economic forces as can be accommodated by the domestic industry without it being swamped. Even the developed countries follow this practice –

giving their industries strong financial and tariff protection in the face of what appears to be more efficient industry in developing countries in some lines of manufacturing. It would be ironic if the developing countries were forced to swallow the full impact of free trade while the developed countries, among them the United States of America, the champion of unfettered world trade, try to shield their industries against imports from other countries.

Case study K

Poor economic policies and state enterprises in Togo

Togo is an African country sandwiched between Ghana and Benin, with a population of 3.1 million in 1986. With an area of only 57,000 km^2, it had a per capita income of US $250 in 1986 and experienced an annual decline in industrial production of about 3 per cent. Manufacturing contributed only 7 per cent to the GDP in 1986; this had declined from 10 per cent in 1965.

When in the mid-1970s, the price of phosphate suddenly multiplied, prosperity seemed to be within grasp for this small country, for phosphate had long been its main export. The phosphate boom gave rise to dreams of industrialization on the western model. Attracted by this new wealth, many European entrepreneurs and crooks appeared on the scene. Corruptibility and the robber-baron mentality complemented each other. But before long the price of phosphate collapsed. Togo's foreign debt exploded and reached US $1 billion. Since 1979, one debt rescheduling has followed another. The profiteers of that time have long since departed with their profits, thanks to state guarantees provided by their own countries. The private creditors have thus been replaced by states that extract interest and loan repayments from Togo. In Togo itself, those who gained exorbitant sums from corruption and were responsible for the wrong decisions are not the ones suffering from the debt burden. It is the population at large and the farmers that pay for the failed industrialization. As a result of the debt crisis, Togo's per capita income shrank by 20 per cent between 1980 and 1984; state services in the area of basic health care, for instance, are now even more inadequate than before. Many of the expensive industrial plants have been shut down or are being sold to

Case study K *(continued)*

foreigners at bargain prices. By the end of 1982, Togo had fallen back into the UN category of the world's 'least developed' countries.

Plastic products made in Togo

A Swiss flag, alongside those of Denmark and Togo, flutters over the grounds of the firm named Industrie Togolaise des Plastiques (ITP) in the capital city of Lomé. The date is 18 April 1986. The Board of Directors is meeting to discuss the company's financial rehabilitation. But the Swiss flag is deceiving – there are no Swiss in the meeting. A few years earlier, it was different. When, on 19 March 1980, ITP was founded, the capital stock of CFA 650 million (US $3.1 million) was held jointly by the state of Togo, Danish sources and the Swiss firm of Promatec of Barestwil. The company's establishment was based on a market study by Promatec, which was also awarded the contract to deliver the turnkey plant and, in the person of Siegfried Weisskopf, provided the firm's first board chairman.

Within the framework of the contract with Promatec, the Swiss firms of Geilinger (Winterthur) took on the steel construction and Buhler (Uzwil) was chosen to supply the machinery. Financing was provided in the form of a loan from the Swiss Credit Bank (Credit Suisse). Without the insurance of this export deal by the Swiss federal government via an export risk guarantee (ERG), the companies involved would hardly have been able to run the risks. On 12 October 1982, bankruptcy proceedings were instituted against Promatec; they were suspended on 15 November 1982, due to lack of assets.

From the beginning ITP operated in the red. By 1984, a loss of roughly CFA 1.3 billion (US $5.1 million) had accumulated. The reasons can be found at three levels:

1 The bureaucratic management style and an excess of personnel are partly to blame. Of the 130 current employees, 35–40 at best will be able to survive the reorganization.
2 The productive capacity of PVC tubes for water pipes, of plastic

Case study K *(continued)*

buckets and plastic chairs was from the start several times too large for the local market, although ITP did benefit from the import prohibition of competing products. But export to neighbouring countries was severely hampered because plastic products are often manufactured locally and, as in Togo, protected from foreign competition. In the view of Yaori Kekey, sales director of ITP, Togo was the victim of an optimistic market study designed to secure the contract for the plant.

3 The price of the turnkey factory was severely inflated. According to Yaori Kekey, the amount charged by Promatec was double or triple the fair market price at the time.

Once the plant had been delivered and after Siegfried Weisskopf had stayed on for a few months as its chairman, the Swiss left and were never seen again. The rehabilitation efforts are now led by the Dutch company Wavin which, together with the German enterprise Pumpen Boese, financed an increase in capital of US $1.2 million. In addition, Dutch and Danish development banks have authorized a 5-year loan for the same amount. Privatization, new management, a heavy infusion of finance capital and large write-offs are designed to help ITP make a fresh start.

Bad investments in abundance

In the case of the plastics factory, as with similar disasters with corrugated iron manufacture and a steel mill, there are connections to Switzerland. But any impression that the Swiss might be acting in a particularly unscrupulous way would be erroneous. Businessmen of other countries have similarly cheated Togo. Even though circumstances will differ from case to case, those described above are nevertheless typical of the manner in which Togo's high indebtedness came about, and they provide the explanation for the financial burdens which Togo suffers today. The foreign loans were used to finance investments whose present real value corresponds to a fraction of the original volume of credits. In a speech in June 1985, the Togolese minister in charge of state enterprises mentioned eighteen firms which, at the time, were closed and

Case study K *(continued)*

awaiting rehabilitation. The textile plants constructed with German support were not functioning at the time of my visit in 1986. The thermal power plant in Lomé, with BBC (a European industrial company) as consortium leader, has been awaiting completion for years. The oil refinery built by the British is closed down – the tanks have been leased to Shell as a regional storage facility. Cimao, a cement factory meant to supply West Africa, financed by French loans and the World Bank, closed after 3 years of operation on 2 April 1984. A misreading of market conditions was the reason for this failure, as well as rentability calculations based on fuel supplies by the Togolese oil refinery. The list is far from being exhaustive. According to a UN report published in 1986 concerning technology transfer to Togo, the extent to which the delivery contracts are supported by a 'multitude of guarantees by the Togolese government' while 'guarantees on the part of the suppliers are almost completely lacking' is 'alarming'. This, it is noted, in no way conforms to normal practice. Corruption within Togo, an aggressive sales policy by foreign business, and state support offered by both the African governments as well as the export-promoting industrialized countries have in their combined effect led to this débâcle.

Reprinted with editing from Richard Gerster (1989) 'How to ruin a country: the case of Togo', *IFDA Dossier* 71.

Key ideas

1 The state plays a crucial role in Third World industrialization, indirectly through incentives and regulations and directly through production of manufactured goods.
2 The present debt situation of many developing countries makes them susceptible to international pressure towards privatizing aspects of their economies.
3 Most developing countries began their industrialization with import substitution involving high levels of protection. However, after rapid early growth, most developing countries began stagnating and turned increasingly towards export-oriented industrialization.

4 The shift towards production for the world market involved a comprehensive review of incentives to favour exports, and the introduction of more competition in the manufacturing sector to equip it for international competition.
5 Export processing zones represent both a spatial unit much like the industrial estate and a comprehensive package of financial and other incentives. The number of export processing zones has doubled in the last decade.
6 Even with pressure to reduce the role of the state in Third World societies, the state will continue to play a crucial role in the future, perhaps more in guiding industrial development than in directly producing manufactured goods.

7
Conclusion

Summary

Industrialization is widely seen in developing countries as the best route to development. Even countries not envisaging a full transformation of their societies aim for partial industrialization, particularly as a complement to agricultural development. Industrialization began in earnest in developing countries only after they gained political independence, in most cases after the Second World War. The only Third World region to begin industrializing much earlier was Latin America, which had decolonized even as other regions were being colonized. Colonialism was not conducive to industrialization since it was in the interest of metropolitan powers to create and perpetuate an international division of labour in which colonies produced raw materials for factories in Europe and represented markets for manufactured goods.

After independence, most developing countries adopted import-substitution industrialization as the quickest route to development. The choice of ISI was based as much on attempts to reduce dependence on developed countries, an understandable reaction given their colonial experience, as on the difficulty of opting for export production, particularly as firms from developed countries had a virtual stranglehold in international trade. Moreover, subcontracting and the general internationalization of production had not really begun in earnest then. ISI afforded these countries an opportunity to develop locally owned industries in a protected environment. However, by its very nature, ISI encountered early problems, such as high production costs because of the reliance on small markets, and lack of innovation due to the absence of competition. In addition, most foreign corporations were interested mainly in establishing branch plants to bypass the protective tariff walls; there was little research and development and no enthusiasm for exports. While some countries, such as Singapore, Taiwan and South Korea, changed their policies in the late 1950s and early 1960s towards export production in an attempt to overcome the problems of ISI, most other countries continued with their ISI policies, and only changed in the 1970s and 1980s. The shift towards export production was necessitated by the major problems of ISI as well as by the historically unparalleled levels of international indebtedness, and the strong pressure of international financial institutions. Currently, export promotion is the favoured industrial strategy, although increasing protectionism in developed countries makes this mode of industrialization risky.

Since the Second World War, there has been a significant increase in the level of Third World industrialization. Currently, developing countries account for about 14 per cent of the world's manufacturing value added. In 1985, manufacturing contributed 18 per cent to the GDP of developing countries (25 per cent in the case of developed market economies and 49 per cent in the case of centrally planned countries). There has been progress, but it has been at the expense of further dependence, and it has been spatially concentrated. The most dramatic industrial progress has been made by a small group of countries, particularly in South-East Asia (Singapore, Taiwan, Hong Kong and South Korea). Other developing countries such as Brazil, China, India, Thailand and Malaysia have also made significant gains in industrialization. However, African countries have experienced economic stagnation, and most low-income countries have made little progress in industrialization.

There appears to be a clear association between the level of incomes of regions and their performance in industrialization, with the already wealthy regions achieving most of the industrial growth. Without significant international intervention in favour of the poor developing countries, industrialization in developing countries is likely to widen further the growing economic gap within the Third World.

Developing countries have succeeded in penetrating the markets of developed countries not only in traditional, labour-intensive products, such as leather goods, textiles and garments, but also in more complex goods, such as electronics products. Some countries are also exporting heavy engineering goods to developed countries, although this is confined to the larger countries such as China, India and Brazil. In general, developing countries rely on markets in developed countries for their industrial products, with only one-third of the products going elsewhere in the Third World. The United States of America is the most important market, followed by the European Community and Japan. Developing countries have also been trying to develop trade among themselves, and have been encouraged in this by the United Nations. However, regional trade blocs have had mixed success only, and developing countries continue to rely overwhelmingly on developed country markets.

A major role has been played by foreign direct investment in Third World industrialization, although this varies very much from country to country. In the last decade, there has been a change in Third World attitude to foreign investment: it is now seen more favourably, and countries are actively wooing multinational corporations. Foreign investment is now increasingly seen as more preferable to international borrowing because repayments in the form of dividend payments are dependent on the performance of the investment; in the case of borrowing, repayments have to be made irrespective of how productive the investment has been. Foreign investment also brings international markets, since most multinationals absorb the output of their Third World subsidiaries within their organizations. Increased foreign ownership of manufacturing in the Third World, however, also poses problems.

A chief means of attracting foreign direct investment is the export processing zone, which has become much more important in the last decade, with the countries possessing them doubling between the mid-1970s and the mid-1980s. This is not to give the impression, however, that locally owned firms are insignificant; they are clearly very important. While the majority of more complex goods are produced by formally registered and organized firms, small informal sector firms have played a major role in Third World industrialization, and most governments have drawn up programmes assisting small-scale entrepreneurs.

Most Third World manufacturing is concentrated in a few urban locations, reflected in the high degree of urban primacy in developing countries (see *The Third World City* by David Drakakis-Smith). The reasons for this are many, including the tendency of the market economy to reinforce initial economic activities and government revenue allocations favouring urban areas. However, many governments are now pursuing policies to decentralize economic activities through selective incentives and regional development initiatives.

The state has been a crucial agent of industrialization in developing countries, not only in regulating and facilitating it, but also in taking part in direct production. Largely because of the relative absence of domestic entrepreneurs, most Third World states became heavily involved in direct production through SOEs. In recent years, however, due to increasing problems with SOEs, which required massive government subsidies, debt-servicing difficulties and pressure from international financial institutions, most states have begun to reduce their role in direct production through a process of privatization. What has become clear from the development experience of the last few decades is that states have to be constantly aware of the opportunities and constraints in the international environment and be flexible in reacting to these opportunities and changes within the framework of clearly laid-out and stable industrial policies. Effective industrial policies have been able to overcome the traditional problems of small market size and poor resource endowment. States have to behave almost like entrepreneurs, except that they have the resources and the responsibility for ensuring that the national interest is protected. In all of this, there is a very important role for competition, but this competition need not be totally open; many of the benefits of competition can be felt through controlled competition, for there is a danger of too heavy losses if the market principle is allowed to run riot.

Recent international experience has also shown the crucial importance of markets in industrialization. Countries with special market access have performed remarkably well in industrialization. To succeed, developing countries will need to continue to seek additional markets, work against protectionist tendencies in developed countries, and cooperate with multinational corporations which offer markets. For small countries, aid and trade agreements, such as the Lomé Convention, provide useful initial market penetration.

Future prospects of Third World industrialization

The future industrial prospects of developing countries depend both on domestic policies and the international economic environment. Domestic policies likely to ensure a better industrial future include the formulation of clear industrial policies; maintaining stability in these policies and avoiding *ad hoc* intervention; reaching a consensus among government, labour and employers on an incomes and labour policy; a commitment to significant competition in the manufacturing sector to ensure that products are internationally competitive in price and quality; and a commitment to the reduction of inequalities among regions and social and ethnic groups.

International issues likely to determine the future prospects of Third World industrialization include the fate of trade talks regarding higher and more secure prices for primary commodities, and on arresting the growing tendency towards greater protectionism in developed country markets. Calls for protection in these countries have increased as a small number of developing countries, such as South Korea, Brazil, Singapore and Taiwan, are seen to have moved beyond being underdeveloped, and pose significant competition for advanced country producers. Furthermore, Taiwan has called attention to itself with a large balance of payments surplus (second largest in the world, after Japan).

The future prospects of Third World industrialization will depend equally on the approach taken on the debt issue because debt servicing represents one of the most significant hindrances to Third World development, making it vulnerable to policies not in its interest, and reducing resources urgently needed for development tasks. To give just one example of the crippling effects of foreign debt repayments: a country could find that in order to repay debts, it cannot import essential raw materials, which leads to a reduction in industrial production. This reduced production in turn means that government revenues, which depend in many cases on sales and customs duties, shrink, forcing the government to reduce its activities affecting the poor section of the population.

The future industrial prospects of developing countries could also be harmed by recent technological developments. Most of these changes involve a major reduction in direct labour costs in production through extreme capitalization (for example, in the use of robots) and through the use of computer-controlled manufacturing allowing shorter production runs and more flexible use of machinery. These also involve a weakening of the labour movement not only to force down wage costs but to ensure more flexible work patterns, including the use of part-time and casual labour. These changes mean that cheap labour, one of the major advantages of Third World manufacturing, could quickly evaporate and manufacturing could once again gravitate towards the developed world. In an environment of free trade, one can foresee this concentration of manufacturing in developed countries, where it will benefit from a more skilled labour force, the location of research and development facilities, and close proximity to markets. This, and the continuing widening of the technological gap, makes the prospects of Third World industrialization rather bleak. Changes in the agricultural sector also cause concern. Recent advances in biotechnology, where living organisms are manipulated in previously undreamed-of ways, is already leading to the production in laboratories in developed countries of crops traditionally grown in developing countries. Given the technological dominance of developed countries, it is most likely that advances in biotechnology will benefit the already developed countries and seriously undermine the development efforts of developing countries. If this were to happen, the whole development potential of developing countries will be threatened.

Even though the trend in developing countries has been towards decreased involvement of the state in manufacturing, its involvement will be critical in the area of technology. Technological choices have a crucial bearing on industrial performance, and governments have an important role in inducing appropriate technology choices. Governments will be crucial in ensuring the effective transfer of technologies as well as in developing indigenous capacity in both technology production and assessment. In particular, developing countries will need to examine whether they need to pass through the same traditional technology paths traversed by developed countries, or whether they should proceed directly to the newest flexible manufacturing technology since their competitiveness is likely to depend on their adoption of these technologies.

The points raised above bring us to a central issue in development studies: there is only one world system and what happens in one part ultimately affects other parts. If developing countries are to make continued gains in industrialization, there will need to be an international effort to assist them, espcially the poorer developing countries.

Developing countries will need to control their rapidly increasing populations too if industrialization and development prospects are to improve. In 1987, 76 per cent of the world's population lived in the Third World, up from 67 per cent in 1950, and projected to rise to 86 per cent by 2100. The rate of population growth is also about three times that of developed countries, with some of the poorest countries recording some of the highest rates of growth. Furthermore, developing countries are also urbanizing at a much faster rate than developed countries.

A reduction in the rate of population growth will mean that instead of national savings being used to meet the social requirements of the additional population, they could be used to improve the economy and assist industrialization. Furthermore, lower unemployment may permit more mechanization, leading to increased productivity and thereby making the manufactures of developing countries more competitive in the international markets.

Finally, the past pattern of industrialization and attitudes to the environment have created a major environmental crisis revolving around the destruction of the ozone layer. The high level of carbon dioxide released into the atmosphere is said to be causing a gradual warming of the Earth, which could lead to the thawing of polar ice and a consequent rise in sea level. If this were to happen, many areas presently cultivated could become uncultivable, and many low-lying areas, including some Pacific Island countries and much of Bangladesh could be submerged. In addition to these global environmental problems, Third World countries will have to address the problems of local environmental damage, such as air pollution and contamination of rivers and lakes. Clearly, then, there is a critical environmental crisis in which industrialization is a key agent.

The environmental crisis affects the Third World most directly in that it would suffer from the consequences more than developed countries because of a low technological and economic ability to deal with the consequences, and because there could be international pressure on developing countries to avoid certain types of manufacturing, even though they have had little to do with the creation of the environmental crisis. This crisis is an international problem and calls for international solutions, but it also calls for recognition that developed countries need to make a greater contribution to solving these problems created by them in the first place. It also calls for a review of our pattern of industrialization so that we can avoid resource-devouring and environmentally polluting types of manufacturing.

Review questions and further reading

Chapter 1

Review questions

1 What links can you see between industrialization and development?
2 What is industrialization? With reference to Figure 1.1, identify the main features of industrialization.
3 Is it true that the international economic environment has influenced Third World industrialization? How?
4 Why is it necessary to understand the different perspectives used in the study of Third World industrialization?
5 List three shortcomings of each of the major perspectives discussed in the chapter.
6 What is ISIC and why is it important?
7 List five types of dependence. How can some of this dependence be reduced?

Further reading

Brookfield, H. C. (1975) *Interdependent Development*, London: Methuen.
Cody, John, Helen Hughes and David Wall (eds) (1980) *Policies for Industrial Progress in Developing Countries*, Oxford and New York: Oxford University Press for the World Bank.
Griffin, Keith (1981) 'Economic development in a changing world', *World Development* 9 (3), 221–6.
Hamilton, F. E. Ian and Hamilton, G. J. R. (eds) (1983) *Spatial Analysis, Industry and the Industrial Environment: Progress in Research and Applications*, Vol. 2, *Regional Economies and Industrial Systems* Chichester: Wiley.
Humphrys, Graham (1990) 'The turbulent world', *Geographical Magazine* 62 (2), 5–6.

Pacione, Michael (ed.) (1985) *Progress in Industrial Geography*, London: Croom Helm.

Sutcliffe, R. B. (1971) *Industry and Underdevelopment*, London: Addison-Wesley.

Chapter 2

Review questions

1 Why is colonialism an important topic for the study of Third World industrialization?
2 Compare and contrast the two main phases of European overseas expansion.
3 What were the main consequences of colonialism for the colonies?
4 Why and how did colonialism discourage industrialization in the Third World?

Further reading

Crow, Ben; Thomas Alan with Robin Jenkins and Judy Kimble (1983) *Third World Atlas*, Milton Keynes: Open University Press.

Dickenson, J. P., Clarke, C. C., Gould, W. T. S., Hodgkiss, A. G., Protheso, R. M., Siddle, D. J., Smith, C. T., Thomas-Hope, E. M. (1983) *A Geography of the Third World*, London and New York: Methuen.

Hill, Christopher (1969) *Reformation to Industrial Revolution*, Pelican Economic History of Britain, Vol. 2, Harmondsworth: Penguin.

Hobsbawm, E. J. (1969) *Industry and Empire*, Pelican Economic History of Britain, Harmondsworth: Penguin.

Myrdal, Gunnar (1971) 'The economic impact of colonialism', in Alan Mountjoy (ed.), *Developing the Underdeveloped Countries: Geographical Readings*, London: Macmillan, pp. 52–7.

Chapter 3

Review questions

1 What level of industrialization has the Third World attained?
2 What is manufacturing value added? What may explain the inclusion of wages in value added?
3 What do we mean by the structure of manufacturing?
4 What are NICs? What explains their relatively better industrialization experience than those of the remaining developing countries?
5 What three generalizations can you make from Table 3.3?
6 Is the South Korean industrialization experience unique?

Further reading

Dicken, Peter (1968) *Global Shift: Industrial Change in a Turbulent World*, London: Harper & Row.

General Agreement on Tariffs and Trade (GATT) (1988) *International Trade, 87–88*, Geneva: GATT.

United Nations Industrial Development Organization (UNIDO) (1988) *Industry and Development Global Report, 1988/89*, Vienna: UNIDO.

World Bank (1987) *World Development Report 1987*, New York: Oxford University Press for the World Bank.

Chapter 4

Review questions

1 Define the concept of informal sector. What role does it play in Third World industrialization?
2 What reasons are often given for supporting small-scale industries?
3 Why is foreign investment crucial to Third World industrialization? What concerns are often voiced about foreign investment?
4 What is privatization?
5 What role should a state have in manufacturing?

Further reading

Rogerson, C. M. and Da Silva, M. (1988) 'From backyard manufacture to factory flat: the industrialisation of South Africa's black townships', *Geography* 73 (3), 255–8.

United Nations Centre on Transnational Corporations (1988) *Transnational Corporations in World Development: Trends and Prospects*, New York: United Nations.

Mazumdar, D. (1976) 'The urban informal sector', *World Development* 4 (8), 120–30.

World Bank (1988) *World Development Report 1988*, New York: Oxford University Press for the World Bank.

Chapter 5

Review questions

1 What is international economic restructuring, and what factors have brought it about?
2 What main factors determine which countries transnational corporations invest in?
3 What factors determine the location of manufacturing within Third World countries?
4 How can we account for the high degree of spatial concentration of manufacturing activities in developing countries?

5 What spatial processes can you observe at work in the case study of Fijian manufacturing location?

Further reading

Boisier, Sergio (1978) 'Location, urban size and industrial productivity: a case study of Brazil', in F. E. Ian Hamilton (ed.), *Contemporary Industrialization: Spatial Analysis and Regional Development*, London: Longman, pp. 182–96.
Dent, D. and Tarrant, J. (1987) 'The old triangle', *The Geographical Magazine* 59, 288–90.
Potter, R. B. and Sinha, Rita (1990) 'NOIDA: a planned industrial township southeast of Delhi', *Geography* 76 (part 1), 63–5.
Smith, David M. (1981) *Industrial Location: An Economic Geographical Analysis*, Chichester: Wiley.

Chapter 6

Review questions

1 What roles does the state play in manufacturing in developing countries?
2 What factors explain the increasing tendency to reduce the role of the state in the manufacturing sector? Should the state drastically reduce its role?
3 What factors have led the state in developing countries to participate in direct industrial production?
4 What is the import-substitution industrialization and what are its main shortcomings as an industrialization strategy?
5 How have governments in developing countries sought to accelerate industrialization?
6 What is an export processing zone and what are some of its main problems as a development strategy?
7 What lessons can we learn about industrial policies from the case study of Togo?

Further reading

Fitzgerald, Randall (1989) 'The privatization revolution', *Economic Impact* 67, 35–45.
Guisinger, Stephen, E. (1980) 'Direct controls in the private sector', in John Cody, Helen Hughes and David Wall (eds), *Policies for Industrial Progress in Developing Countries*, Oxford; Oxford University Press, pp. 189–209.
Hardhill, Irene (1986) 'The Shenzhen experiment', *Geography* 71 (2), 146–8.
Lal, Deepak (1980) 'Public enterprises', in John Cody, Helen Hughes and David Wall (eds), *Policies for Industrial Progress in Developing Countries* Oxford and New York: Oxford University Press for the World Bank, pp. 211–34.

Index